M000158323

Standards
of Brewing

A Practical Approach to
Consistency and Excellence

Charles W. Bamforth, Ph.D., D.Sc.

 BrewersPublications.com

Brewers Publications
A Division of the Brewers Association
PO Box 1679, Boulder, CO 80306-1679
Telephone: (303) 447-0816
BrewersAssociation.org • BrewersPublications.com

Printed in the United States of America.

10 9 8 7 6

ISBN 13: 978-0-937381-79-3
ISBN 10: 0-937381-79-9

Library of Congress Cataloging-in-Publication Data
Bamforth, Charles W., 1952-
  Standards of brewing : a practical approach to consistency and
excellence / by Charles W. Bamforth.
      p. cm.
Includes bibliographical references and index.
  ISBN 0-937381-79-9 (alk. paper)
  1. Brewing--Quality control. 2. Beer. I. Title.
  TP570 .B185 2002
  663'.3--dc21
                                        2002008675

Technical Editor: Inge Russell, Ph.D., D.Sc.
Copy Editor: Elizabeth Gold
Cover and Interior Designer: Julie Korowotny

This book is in tribute to two giants of quality
in brewing: Geoff Buckee and Harry White

To Diani

[signature]

Contents

Acknowledgements

I thank Toni Knapp and Ray Daniels for their "strong" words of encouragement when the muse wasn't with me and their "gentle" suggestions concerning deadlines. I would like you, the reader, to prove Toni wrong about one thing: she said this isn't bedtime reading. I insist that it certainly is something to cuddle up with—but you will need a calculator in your jim-jams! I am reminded of the time I told my boss, Gus Guthrie, that I had Stephen Hawking's, *A Brief History of Time*, by the bed to impress my wife. "Why, what do you do?" he asked, "stand on it?" Hopefully reading the present tome would be the more credible option.

Ray Daniels provided a great deal of useful opinion on the clarity with which really quite complex concepts were being positioned. The entire text was meticulously checked by Dr. Inge Russell.

Grateful thanks to Ken Grossman, owner of Sierra Nevada Brewing Company, for generously contributing the Foreword.

And as ever, thanks and love to my patient wife, Diane. Bennatew genough.

Foreword ━━━━━━━━━━━━━

By Ken Grossman
President, Sierra Nevada Brewing Company
Chico, California

Whether you operate or work in one of the world's smallest breweries, or in a brewery that produces millions of barrels per year, a sound and practical quality assurance program is a must for long-term survival in the brewing industry today. The marketplace demands that are now placed on breweries to produce and deliver consistent, fresh tasting beer has never been greater. Consumer fascination with novelty alone has passed. Breweries will not be able to survive just because their beer tastes different than the mainstream. Whatever the style, the consumer will expect a specific beer to taste the same every time they make a purchase. If not, they now have plenty of alternatives.

Many of the breweries that were established in the early days of the American craft beer resurgence are gone and many more are having problems. You could point to many possible reasons for their failure, such as lack of capitol, access to market or poor business skills, but product quality has probably been the biggest downfall for most of these breweries. The brewers that survive and prosper going forward will have to be resourceful, energetic, a little bit lucky and above all else consistent producers of great beer.

The challenge facing all breweries is establishing a sensible program that satisfies the sometimes conflicting goals of maintaining quality assurance parameters while meeting real life production and sales demands. The very nature of the brewing process—from the ever changing raw materials and easily influenced biological processes to the complexities of packaging and distribution—necessitates that the brewer develop an arsenal of tests and procedures that will help insure the consistency and stability of the beer in the marketplace. Obviously the needs and demands of the QA program for a 400 bbl a year brewpub, selling only on premise, and the 100,000 bbl national distribution brewery are going to be significantly different. But establishing and prioritizing an effective QA program will be equally critical for the success of both operations.

I feel it is essential that the top management and decision makers in the brewery understand and promote the never-ending drive to improve and maintain quality. This vision has to be instilled in every employee who works in your brewery and I feel this is only possible if the management takes it seriously. I also think that you need to have some level of technical expertise at the helm of a brewery. The people who control the purse strings need to understand why they need to continually invest in improved technology. It may be impossible to convince a non-technical person that it's a good investment to spend a seemingly outlandish amount of money on a modern bottle filler when the old one is still capable of filling bottles, although they may be unwittingly damaging their company's future. Over the years I have also been amazed at the short sightedness of some brewers, either unaware, or worse knowingly, selling seriously flawed beer. I guess they assume most consumers can't tell the difference, or possibly they can't tell themselves, either way they are helping to write their companies' obituaries. Not that any brewer ever wants to dump a batch of beer, but if it is seriously flawed, don't compound your problem by selling it or blending it with good beer, use it as a learning experience. Even though it may be financially painful to dump it, you may receive some value by showing the staff your commitment to quality.

Dr. Charlie Bamforth is in a very unique position to write this book, having worked both as a brewery quality manager and as a researcher and academic in the field. He has first-hand experience of the challenges and pitfalls that come with establishing and administering an effective QA program. Charlie understands the role that every department must play in maintaining quality parameters in cooperation with the QA team. His focus is on establishing a practical program that utilizes common sense methods, as well as setting up systems to prevent many potential problems, rather than on finding them after it's too late. Equally important he tries to put in perspective the need to validate the data and methods upon which these critical decisions are based. Charlie also conveys a keen sense for the balance that the brewer must maintain between ideal specifications and the realities of available raw materials and process variables. Although this book is written primarily from a technical standpoint, I think everyone who works in a brewery will find useful insight.

As quality assurance manager of a major brewery in the northwest of England, I had the task of opening letters of complaint. Each week they would come in, matched by reports from our pubs and trade accounts of products that were less than perfect.

Somehow we once contrived to color up a lager and accidentally produced a beer that, by comparison, made Guinness look anemic. Disastrously, the beer somehow got out to trade and was met with numerous complaints. Curiously, however, not everyone complained.

Occasionally there were reports of things floating in our beer. It's amazing how often bits of glass contrived to find their way into cans, usually just after newspapers reported slithers of glass being found in jars of baby food. Sometimes such complaints were genuine, even down to the discovery of a can containing a condom. In another instance, a fellow QA manager claimed to have received two bottles in successive mail deliveries: the first contained the front half of a mouse, and the second contained the back half, tail included. I think he was pulling my leg, though I can't be certain.

Rather few and far between were moans about the taste of our beer. Most often people protested about the foam. Even though one is never totally sure whether some beers are supposed to taste the way they do, any idiot can tell whether the beer has a head on it or not.

I reckon we got about 20 to 25 complaints each month, about 300 in a year. (Incidentally, if that math is beyond you, you probably shouldn't be reading this book.) That's 300 complaints for a brewery putting out more than 1 million barrels of beer annually. That's a U.K. barrel, by the way, which is 20 percent bigger than a U.S. barrel, even though I was brought up to believe that everything is bigger in the states, especially in Texas. If we put these figures into can terms, each of those cans held half a liter of beer, amounting to more than 325 million cans. Roughly, that equals one complaint for every million cans.

When you look at it that way, it's not very much. In fact, I reckoned my efforts meant I was worth an extra zero on the salary check each month. That, however, is not the essential point I seek to make here. Instead, I want to use this data to emphasize the remarkable consistency of the brewing process. Minute after minute, hour by hour, day after day, the packaging lines of breweries are churning out beer. Some of the canning lines are running at more than 2,000 cans every minute. The beer in those containers is astonishingly consistent. It is consistent because brewers care about, and have gained a wonderful understanding of, their raw materials and processes. The knowledge enables them to tweak this or twiddle that so that every drop of beer tastes the same and looks exactly as is intended.

I have some good buddies in the wine business here in northern California. Wine is a fine and noble product. Unlike some winemakers, though, brewers don't fall back on the safety net of "vintage." I cannot imagine the brewmaster at a brewery of one of the world's major companies responding to the glare of the top man from HQ with "Yep, I know it's not good, but give us a break, the hop crop this year has been lousy."

Brewers have just as much to contend with as winemakers with regard to the vagaries of the relevant crops, in this case primarily barley and hops. Season by season, variety by variety, growth location by growth location, crops can be enormously different. The maltster and the brewer respond to these challenges by adjusting their processes, making the best of what sometimes is a pretty bad job.

In order to do this, they need information. They need numbers. They need something they can use to make a judgment on the raw materials, the process, and the product— to decide whether it's a go or not.

That is what this book is all about. Its purpose is to describe (in as friendly a manner as possible) the diversity of tests that are applied between the growing of barley and the beer at the point of sale. I strive to demonstrate what the numbers mean and how they can be interpreted.

This book builds on a class I teach at the University of California, Davis. In it we instruct the students on how best to brew a beer and how to analyze raw materials, process streams, and the finished product. They learn how to interpret specification sheets, how to respond to samples that look a touch shabby, how to troubleshoot problems—even disasters.

This is not a brewing textbook. If it's basic brewing science you're after, look no further than the book my friend and predecessor, Michael Lewis, originator of the class at Davis, wrote with Tom Young. (See Bibliography for a recommended reading list. Appendix 2 should also help the reader who is totally new to the world of brewing. And go to Appendix 3 for a reminder of the beauty of chemistry!) This book is neither a recipe book nor an operations manual. Instead, it aims to capture the practical day-to-day reality of what quality assurance in the maltings and brewery should be all about.

This information should prove valuable to students of brewing by giving them the opportunity to hone their investigational and interpretative skills. I end each chapter with a series of exercises similar to the quizzes and examinations I give my students. They love the challenge—I hope you do, too!

Chapter One ▄▄▄▄▄▄▄▄▄▄▄▄▄▄▄▄▄▄▄▄▄
The Brewing Technician

I was sitting next to the company chairman once at a fancy dinner. He was actually a kindly man, albeit a chap one was naturally wary of, as is generally the case for the breed. I don't recall the gist of what I said during the course of the evening but I do remember something I said right at the end of a reasonably convivial gathering.

"Mr. D.," I said, "I am just a simple scientist."

He looked at me, smiled enigmatically, and replied, "Yes, Charlie, I think you're probably right." I didn't take offense. Basically what he was implying was that I was not really cut out for the ruthless world of business dynamics, higher management politics or financial subtleties. Rather, I was from a world where the prime currency was scientific intellect, the *raison d'être* was technical excellence, the beauty was in the consistency of the beer and not the bottom line of a set of accounts.

I wasn't so naïve as to ignore the importance of profit margins, stock performance, and the like. Most important for me, though, was the fact that Mr. D. and his main board colleagues were not stupid enough to overlook the fact that the success of the company at the time was solidly based on a foundation of superior practical performance in our breweries.

"There are two things that are of the top priority in our company, Charlie," said Robin, another first-rate member of the same board. "The first is our people, the second is quality. And if we look after the former, they will look after the latter."

I considered myself fortunate to be employed by such a technically- and quality-driven company—ethics that placed them at the very top of the brewing tree on our patch. There were two technical men on the main board. Folks I knew at other breweries weren't as fortunate as I was because "bean counters" were at their helm. They were led by people for whom the bottom line meant everything and who would cut all manner of corners in their pursuit of a fast buck.

Even today I can take you to several types of brewing companies. There are those for whom happiness is the pursuit of the quick pint and who openly profess their ambition to brew beer in a day. Rather closer to my own ethic are those who realize that quality will win out in the end. They follow the idea that, if it ain't broke, then for goodness sake don't try to fix it. The fact is that the actual cost of the ingredients in a glass of beer is really rather low. The act of putting beer into a can or bottle tends to cost substantially more than the liquid itself. It's the cost of producing the beer (i.e., people), the call-off by the taxman, and the ever-growing costs of sales and marketing that slap the mark-up on the product.

One of the classic examples of the folly of futurism came a number of years ago with a famous name in North American brewing. The company employed an impressive array of technical talent, guys perhaps driven more by the satisfaction of leading the innovation charge than by ensuring as the number one priority that nothing, but nothing, deflected them from the over-riding goal of quality and consistency. The company proudly boasted of its developments in the area of fermentation and how it could really accelerate the process with adjustments such as stirring the contents of the fermenters. Its misfortune was that, no sooner had it bragged in very public circles about its achievements, calamity befell the company in a way that had nothing to do with the speeded up process but everything to do with ignoring the basics of quality. The company developed a problem with particles ("bits") in the beer of snowstorm proportions, due to the injudicious use of stabilizers. There was no way that the stock market would disconnect the two things and therefore the market deduced that here was a company that lacked the necessary robustness to manage its own business. Result: defunct brewer.

There are various messages in this story, the overwhelming one of which is: look after your beer. As a scientist and no Luddite, I certainly won't deny the need for, and the benefits to be had from, carefully controlled research and development. The brewing industry today would be in a sorry mess if it was not for the development of the technologies that you see about you in any properly run and organized brewery. However, technological change should be evolutionary rather than revolutionary. In the superb company I used to work in, we were pretty much given free rein to vent our wacky spleens within the lab and in the experimental brewery. As soon as the brewery got scent of us, however, the commonsense police kicked in and every control imaginable was put in place to ensure that nobody and nothing screwed up the beer that was everybody's bread and butter.

The selfsame precautions should be taken against the bean counters. I can take you to a certain brewing company today where the worm has turned full circle. The breweries are filthy. Why? Because it costs money to mop up the spilled sugar. The beer has been thinned down so as to contain less alcohol. Why? Because it saves on the tax bill. And so on. The folks are demoralized. The incentive for quality has vanished.

Whether you are ankle deep in shag pile carpet in a penthouse office or the guy checking the alignment of labels in the bottling hall, your overriding ethos has got to be quality. It should be agenda item *numero uno* at the management meeting. Quality is the fulcrum about which the entire success or failure of a company stands.

One of my favorite quality stories is of the company that set a specification for some gismo or other and made an agreement with a supplier in Japan. The demand was that 98 percent of the product should fulfill the target requirements. Days later, two packages arrived: one very large, another much smaller. A note accompanied them. "Please find a large container containing those gismos within specification and a second package containing those outside specifications. We are curious about what you are going to do with the latter."

Probably apocryphal. But amongst the lessons to be learned is one of meaningful specification setting and agreement. This is at the very heart of the achievement of quality and consistency. Realistic specifications need to be agreed, for instance, between maltster and brewer, between hop supplier and brewer. Appropriate specifications throughout the brewery so that those operating the process at each stage—brew house, cellar, packaging—know what they are dealing with.

And make the specifications ones that will mean something to those at the top of the company. They should understand as well as anybody else what it means and how to know if their beer is drifting high on diacetyl; if their foam stability is down the pan; if there is mycotoxin in the malt; if the yeast is displaying curious tendencies.

There are several companies in North America that are shining examples of success in response to having people at the very apex of the organization with a finely tuned technical appreciation and feel for quality. The world's biggest brewer by far, Anheuser-Busch, is driven on an ethos of absolute quality and consistency. On a different scale, Sierra Nevada, out of Chico, has grown in 25 years from converted dairy equipment to a 500,000-plus barrel per year trans-continental operation on the framework of getting it right.

Methods. Specifications. Understanding. Application. That is what this book is all about.

Chapter Two ▬▬▬▬▬▬▬▬▬▬▬
Principles of Quality

There are plenty of definitions of quality. A favorite one is "the supply of goods that do not come back to customers who do." Another is "the extent of correspondence between expectation and realization." Simply put, that means, "the match between what you want and what you get."

What constitutes quality for one person may represent quite the opposite for another. I am an Englishman. For me, peanut butter and jelly sandwiches are not synonymous with quality dining. By the same token, many people reading this book would probably distance themselves from my notion of perfection in the culinary world, namely Onion Bhaji's and Lamb Pasanda washed down with several pints of lager.

Limiting ourselves to the world of beer, we must accept that one bloke's nectar is another's poison. Personally I abhor many of the hefeweissens dispensed in brewpubs here in California. For me, most quality beers are bright. Similarly I rail against lightstruck character in beers—there really is the most remarkable and awful convergence between this aroma and the smell of squashed skunk on the freeway. And yet there are ample people who seem to mind neither that their beer looks like chicken broth nor that they are savoring a product that reeks of perineal outpourings. To me a low carbonation ale (not nitrogenated), with well-balanced dry hop notes, dispensed with a modicum of foam approaches perfection on a cold winter's eve in Sussex. Many folks would share that view of quality (including the Campaign for Real Ale), but others would not. Who is to say who is right? It's horses for courses. Equally, if I am watching the Sacramento River Cats, it's 104 °F (40 °C) and I'm stuffing nachos, a gently-flavored American lager is sublime. The aforementioned ale would not fit.

Let us simply accept (and rejoice in the fact) that there is a rich diversity of beer types. I suggest that the mission of any brewer must be to ensure that whatever the genre under production, it should adhere to this principle criterion of quality: quality is the achievement of consistency and the elimination of unwanted surprises.

Them and Us
When I was a quality assurance manager, I felt very much a second-class citizen to the head brewer, even though we were equals on the management team. For starters, he was the guy to get slushed—fancy meals paid for by maltsters and hop suppliers, bottles of whisky from grateful suppliers, and so on. More importantly, he was the bloke who ended up taking the key decisions. I was the irritant, the bloke whose team had the responsibility of policing the system and, rather like a classroom snitch, I was the person expected to creep to the plant director with tales of how the production folks were screwing up.

Perhaps I'm exaggerating, but the "them and us" mentality really is all too prevalent in many companies. Quality is not the exclusive preserve of a quality control, or even of a quality assurance department. Rather, quality depends on a commitment from all employees.

It is fashionable today to speak of Total Quality Management (TQM). Everybody in a company must have a commitment to excellence. It doesn't matter whether they are on the production line, in the lab, performing janitorial functions, or masterminding the whole shooting match from the comfy swivel chair. Everybody must realize that their input will impact the output from the company and will contribute to whether it is a production with genuine quality or not.

To attain a desired output, quality must be taken into consideration throughout every facet of a company's operations: design, manufacture, marketing, purchasing, etc. It needs to be driven from the top. All employees should realize and recognize the commitment that senior management has to achieving a quality delivery, for only then will the necessary momentum towards an all-embracing quality operation be possible.

Quality Systems

TQM is, in part, achieved via the adoption of a quality system. Nowadays formalized approaches are available, embodied in international standards such as those of the ISO 9000 series. These are exercises in focusing the mind, achieving compliance, and ensuring that systems are documented. They represent good discipline. However it must be realized that they don't necessarily ensure that a product is good or bad. They don't guarantee that process stages are necessarily the correct ones. All they seek to do is ensure that standard procedures are followed. It is up to the company to ensure that best practices prevail. To have the ISO 9000 accreditation is really a stamp of approval that a company is paying heed to the need for quality systems.

Some companies actively seek out suppliers who possess this type of accreditation. It is far sighted for companies to look for suppliers that have genuine quality systems in place because that means there is presence of, and adherence to, a quality manual. Such a document should be gospel in any company, brewer and supplier alike. The manual should document everything that pertains to the product: specifications for product, raw materials and process streams; procedures for everything from materials purchase, storage, handling and use, throughout the brewery to shipment of the beer.

Quality Assurance Versus Quality Control

The traditional approach is to have two separate bodies (bringing with them, regrettably, a degree of mutual distrust and distaste). The production guys make the stuff, and the quality control folk pull it apart with the aid of countless measurements and say what does and what does not conform to specification. Quality Control (QC) is a reactive approach. The serious shortcoming of this system is that it can be associated with waste: it is simply not good enough after the event to pick and choose what is and what is not able to go to trade. Or, even worse, to identify something as being less than ideal and then letting it go to trade anyway. This takes me back to the time that (as QA Manager) I put a hold on a bottled ale owing to an oxygen content over specification. The warehouse guys were jumping with rage. "We need that beer," they screamed at the managing director, apparently there being a shortfall in the trade. I was persuaded to rescind the hold. Logic kicked in: if there was such a short-term emergency need, I rationalized that the beer would be supped before it had gone stale. However I was grateful that I belonged to an organization where the QA manager was not beholden to the production people directly in a line reporting structure, for then I would not have been in a position even to delay matters such that the situation could be discussed from a rational perspective.

Much more effective is to establish a Quality Assurance (QA) approach in which systems are introduced that ensure that at every stage in its production, the product is within specification. The emphasis is one of prevention rather than detection. The buzz phrase is "Right First Time." This avoids the need to make retrospective decisions about whether an out-of-specification product is or is not able to be released.

Some analysis and quantification of the process and product are naturally essential. Wherever possible, those measurements should be ahead of a critical event rather than after it. So much better to assess the yeast pitching rate and required oxygen content for a specific wort at a given temperature and thus ensure a controlled and predictable fermentation than to faff about trying to stimulate sluggish fermentations or blend away lousy flavors. Far wiser to taste the water on a go/no go basis than end up wondering what to do with a warehouse full of tainted beer.

It should be self-evident that the most appropriate people to make the assessments are the ones who are operating each stage of the process. Wherever possible, measurements should be made automatically by some sort of in-line sensor that is able to trigger a feedback loop. (See Figure 2.1) Another option is for the

Figure 2.1
An Example of a Feedback Loop

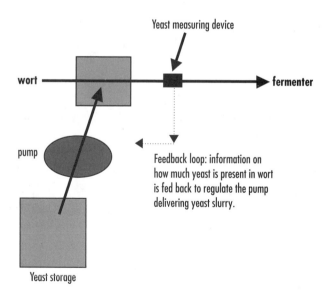

The amount of yeast present in the pitched beer is quantified, with the value feeding back (preferably automatically but, if not, manually) to the pump regulating the dosing of yeast slurry.

operator herself to take ownership of the issue and avoid being hassled and irritated by some sprog in a white coat from the lab, professing disingenuously that he's only there to help.

Designed-in Quality Versus Conformance

It is important to differentiate between two distinct aspects of quality. The first, which I call designed-in quality, is a measure of whether a product is designed to meet its purpose. We are really talking about specification. If the specification is adrift of what is appropriate, then the desired quality will not be there. Let's say, for instance, that our concept of quality is canned beer that will not display stale character in six months of optimum storage. In that case, setting a specification of 0.5 ppm for the oxygen level in the product as it goes to trade would be dumb. We should be looking at a tenth of that value or better.

Another element of designed-in quality concerns the specifications we set on our raw materials. If we want our beer to have a dimethyl sulfide level of 50 ppb, it would make little sense to leave dimethyl sulfide precursor off the specification list for the malt. It would be stupid to set it at too low a level.

The more we design-in quality at the various process stages—from raw materials onwards—the greater the likelihood of achieving excellence in the end product. I must stress, though, that the parameters that are specified should be relevant and attainable. It certainly does not make sense to overgild the lily by worrying about measures that have no bearing on the final outcome. This is especially true if the measurement and the re-work they would cause if they were out of specification would represent an extra cost burden.

The "conformance" element is a measure of the extent to which the product satisfies the designed-in quality. It is simply an index of how successful the quality assurance approach has been. Conformance measurement is QC as opposed to QA. Its purpose should mainly be to allow a tweaking and an adjustment of the

Figure 2.2
QA versus QC

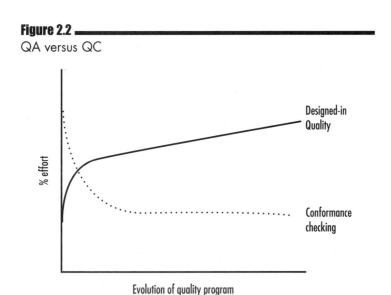

Evolution of quality program

The relative effort of ensuring conformance through "designed-in" quality (QA) and measuring conformance (QC) as a quality system evolves.

designing-in exercises. Ultimately the ratio of designed-in to conformance elements should rise to a plateau. (See Figure 2.2)

The Role of the QA Team

The principle function of the QA manager and his team is to maintain the quality manual. This manual is the document that lists product recipes and specifications, documents who does what and when, tabulates raw material tolerances, explains troubleshooting protocols, etc. I must stress that the production of this manual is the responsibility of everybody in the operation, who must not have it foisted upon them by the QA department. Everybody needs to take ownership. The QA team merely ensures that it is kept up to date. Procedures and specs may change, in response to continual improvement, new product lines, equipment replacement etc.

The Cost of Quality

There is no better way to illustrate the merits of a quality system than to quantify it in cost terms. We can divide quality-related costs into several categories.

Internal Failure Costs

These are associated with the product being out of specification at whatever stage in the production process and before it has passed to the consumer. We can subdivide into:

- Rework—the cost of correcting matters to return a product to 'normal.' An example might be the implication of a beer arriving in the bright beer tank after filtration containing too much oxygen. The tank would need then to be purged with CO_2 or N_2 to drive off the oxygen—a practice in which foam is often produced. This tends to collapse to form bits, demanding re-filtration if the bits are not going to detract from final product quality.
- Re-inspection—the cost of checking re-worked beer
- Scrap—product that is beyond repair. An example would be gushing beer. If the beer has gotten into the final container and displays gushing, there is nothing you can do but destroy it. It's important to ensure that good and wholesome malt was used in the first place.
- Downgrading—product that can be traded but only in a down market way—i.e. sold off to somebody to make into vinegar
- Analysis of failure—the cost of getting to the bottom of and rectifying the cause of the internal failure

External Failure Costs

These measure the implication of a product actually getting out into trade with a quality defect:

- Recall—the cost of investigating problems, recovering product and, probably above all else, the cost of lost reputation and market. An example would be the "bits" problem ignored by a major U.S. brewer, which as a result became a former major U.S. brewer.

- Warranty—the cost of replacing product
- Complaints—the cost of handling customer objections
- Liability—the implications of litigation

The two categories of failure costs are the upshot of quality failure. The other element of quality costs are the costs associated with getting it right.

Appraisal Costs

This is the price of analysis throughout the process, from raw materials to end product, and can be sub-divided into:

- Inspection and test, namely the costs of getting the analysis done but also the cost of calibrating equipment and sundry other costs involved in running a lab
- The cost of internal auditing
- The cost of assessing suppliers (audits, approvals)

Prevention Costs

This is the expense associated with people operating the quality system and is basically the costs devoted to operating a well-run QA system, one that is set up to ensure that the product is right first time all the way through the plant. (See Table 2.1)

Getting Quality Costs in Balance

It clearly makes sense to invest in prevention costs provided it enables a decrease in appraisal costs and a decrease in failure costs. (It really shouldn't be necessary in a properly trusted system to insist on blindly continuing to measure everything in sight.) However there comes a point when a continuing investment in prevention costs is no longer beneficial and means nothing more than a financial liability. (See Figure 2.3)

Figure 2.3 ━━━━━━━━━━━━━━━━━━━━━━━━━━━
Quality Costs as a Quality System Evolves

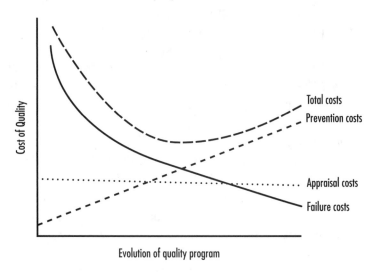

Evolution of quality program

Converting This Into Money

It has been estimated that the typical cost of quality in a business such as brewing is 5 to 25 percent of the total cost of sales. Let's take the low end of this range for a brewery with an income of $100 million and a profit of $10 million. The cost of quality is therefore 5 percent of $90 million or $4.5 million. You will recognize that fluctuations of this cost, either upwards or downwards, can be relatively large in relation to the profit margin. Attention to these costs can make a major contribution to the bottom line. Equally, inattention to matters of quality could wipe out profits rather too readily for comfort.

Perhaps if some of the captains of our brewing industry better appreciated the true cost of quality in actual dollar terms, it might hone their interest in this area. It is not a straightforward task, but a good starting point would be to convert certain performance criteria into manpower costs and costs of materials. Relevant data would come from:

- Raw material usage
- Brew house yields
- Plant utilization and efficiency (i.e., on packaging lines)
- Re-processings
- QC activity
- Beer losses
- Trade returns
- Trade quality

Quality Is a Shared Responsibility

Once generated, the quality cost information should be shared as widely as possible through a company. What better illustration could there be for employees about how their role feeds into the overall quality mission of a company? It's also helpful if quality performance is built in to the performance appraisal for employees. Substantial improvements in the quality cost balance must reflect the role of all parties relevant to a drive towards quality.

Table 2.1

Quality Assurance Department Roles

Function	Comments
Coordinating the quality system: what is worth measuring, when, where, how often and by whom	Note: I stress "coordinate." All parties should buy in to this. The QA folks can advise on issues such as statistical relevance; the brewers should chip in with expertise about relevance. The engineers should comment on plant design. Interactive discussion should lead to the best balance of process efficiency and quality.

Function	Comments
Establishing the quality standards and procedures for raw materials and spot-checking them	Agreement with the supplier is essential. Specifications should be fair and achievable.
Auditing and surveying of processes and procedures	Audit teams should ideally comprise a mix of personnel from different functions—"buy in."
Projects	E.g., installation of in-line systems
Training and auditing of operators on QC checks	Ensuring that those making the measurements at line are performing the methods correctly
Collaborative exercises	Participating in inter-collaboratives between labs and organizing those between satellite labs within a brewery
Specialized analysis and procedures	E.g., gas chromatography, high performance liquid chromatography, formalized taste tests
Troubleshooting	Again should have buy in from production personnel—"mixed" teams
Product surveys	Products within the brewery versus those from outside
Complaints	Improving procedures in response to them
Quality awareness campaigns	Posters, videos, quality circles (discussion groups), etc.
Instrument checking and calibration	Ensuring that all lab and on-line equipment is reliable
Coordination of quality systems	E.g., ISO 9000, maintaining the quality manual
Coordination of cost of quality information	Making available control charts that illustrate quality issues on a direct data but also on a $$$ basis

Chapter Three
Statistics and Process Control

It was Benjamin Disraeli, prime minister to Queen Victoria, who said, "There are three kinds of lies: lies, damned lies, and statistics."

Personally, I prefer the thesis of Aaron Levenstein, whose opinion is that statistics are like bikinis. What they reveal is suggestive but what they conceal is vital.

However, if we are to talk sensibly about the analysis of beer, of the materials that go into the making of beer, and of the process stream that leads to beer, we can't stray very far from our statistics textbook, at the very least Statistics 101. We must have confidence in what our instruments are telling us. We must know whether the differences that we see between samples are meaningful and significant.

Some of the methods I will refer to in this book are relatively subjective and depend on the opinion of one or a very few people. For example, the taste screening of brewery waters is likely to involve at most a roundtable of people (brewers, quality assurance staff) striving to ensure that no taints can be detected. Such is frequently the case for the tasting of beer. We will encounter relatively sophisticated tasting protocols that rely absolutely on statistical interpretation and are essential for dispassionate conclusions to be drawn in many instances. Nonetheless the executive palate holds in many locations. The fatter the paycheck and cigar, the more acute, skilled and authoritative seems to be the palate, giving greater weighting to the result. A senior exec's "two" is likely to be bigger than a "four" from the proletariat.

I recall (as a headquarters guy there to help) discussing a lousy beer with the managing director of one of our breweries. The beer had an overwhelming butterscotch aroma, and the measured diacetyl figure was through the roof. We in HQ didn't like it one bit. He tasted it, looked at me, and in a state of mind and body that would have passed any polygraph test, said coldly, "This beer has no defects."

Many methods are more objective than this. They are open to critical statistical examination.

Vital Statistics

I hated mathematics as a schoolboy. It was probably because of a teacher who daily had me stand on my feet ("On your hind legs, boy!") and stagger through what to me seemed not so much Greek as Venusian. I somehow passed the exams but I was left forever more with a suspicion of matters numerate.

Which makes it all the stranger that I am writing this book. Fact is, you might not enjoy the math and statistics but you can't deny their importance to the achievement of a properly controlled process. Ergo, if you can't beat 'em, join 'em. Which leads me to the fundamentals of the use of statistics in brewery quality control.

Some Definitions

It is important to have tools for locating the various measurements we make so we can position them in relation to others. We also need to quantify how disperse the data is. How broad is the spread?

To position the data, we need some index of where the middle of the data is positioned. There are three ways of doing this, by looking at the mean, the median and the mode.

The mean is basically the average of the values and is obtained by adding them up and dividing by the number of measurements.

Example: The mean of the fifteen measurements 1.3, 1.9, 1.4, 1.2, 2.8, 1.3, 1.4, 1.4, 1.4, 1.4, 1.1, 1.3, 1.7, 1.8, 1.4 is

$$\frac{\begin{array}{c}1.3 + 1.9 + 1.4 + 1.2 + 2.8 + 1.3 + 1.4 + 1.4 \\ + 1.4 + 1.4 + 1.1 + 1.3 + 1.7 + 1.8 + 1.4\end{array}}{15} = 1.52$$

Remember algebra? We can describe this sort of calculation in algebraic terms. Call the various measurements made x_1, x_2, x_3 etc (i.e., in this example x_1 is 1.3, x_2 is 1.9, and so on). Let's call the total number of measurements made n (i.e., here n is 15). Let's call the type of value i. The mean can be called M, and the total number of measurements made (i.e., fifteen) we'll call Σ (big sigma, you'll recall). If you look at what we just did in the calculation example, it can be described by:

$$M = \frac{1}{n} \Sigma x_i$$

We added all the values together and divided by the total number of those values.

The *median* is the value bang in the middle of the data set when you arrange them in ascending order. To take our data set just figured and arrange it in ascending order, we get 1.1, 1.2, 1.3, 1.3, 1.3, 1.4, 1.4, 1.4, 1.4, 1.4, 1.4, 1.7, 1.8, 1.9, 2.8. The median is 1.4. If the data set comprises an even number of values (say it had been 16 instead of 15), it is conventional to take the midpoint between the middle values (in this case, that would be the eighth and ninth points). In the present example the value would still be 1.4. However say that the 8th and 9th values had been 1.4 and 1.5, the median would have been 1.45.

You will see by comparison of these mean and median values that the latter gives a closer indication of the values that predominate (i.e., at the lower end of the range). The mean is higher because there are one or two values that are quite high, presumably atypically (i.e., 2.8).

The third way of indicating location is the mode. This is simply the most frequently occurring value in a sample set. In the present example, the value is 1.4. If all 15 numbers had been different, we would not be able to define a mode. Mode is seldom useful as a term.

Data Spread

Now let's consider how we can describe the extent of spread (dispersion) of the data. In our example, we obtained a mean of 1.52. If all of the 15 values had been 1.52, we would have exactly the same mean (and the identical value for median and mode). But the 1.52 we did get is the average of a much greater spread of numbers. Clearly a value such as the mean does not give us sufficient information.

The obvious value to quote is the range. In our example, this is 1.1 to 2.8. It tells us how widely spread the data set is, but of course it tells us nothing about where the data is congregated. It doesn't tell us that the value of 2.8 is way out. Without seeing the actual 15 data points, it's only if we quote range and mean and median that we can conclude that most of the data must be congregated closer to the lower end of the range.

Of course if we plot the data we can visualize this very readily. The plots might be in the form of histograms (Figure 3.1) or curves (Figure 3.2). You can see that there may be various shapes for these plots, illustrative of the range of data we have. Mathematicians, though, love their algebra. What algebraic terms are available for describing range?

For any individual measurement, we can describe how much it deviates from the mean by the expression $x_i - M$. For our example, the value 2.8 represents a deviation of (2.8 minus 1.52) or 1.28. The value 1.1 represents a deviation of -0.42. If

Figure 3.1 ━━━━━━━━━━━━━━━━━━━━━━━━━━━━━━━━━
Data Plotted as Histograms

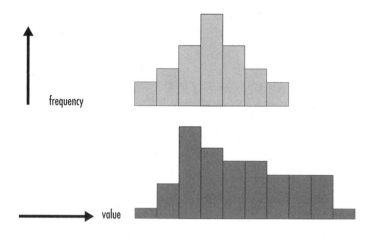

Figure 3.2 ━━━━━━━━━━━━━━━━━━━━━━━━━━━━━━━━━
Data Plotted as Curves

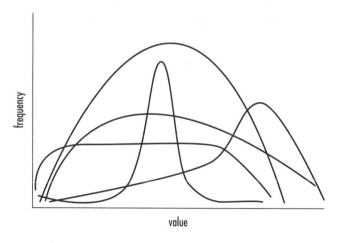

we add all the deviations together we will obviously end up with zero. Zero is seldom a very valuable number unless it's something like the number of penalty points on a driver's license or the number of goals scored by the opposition. However if we ignore the + and − signs on the deviation, we can get a useful measure of the extent to which the data deviates from the mean. The value is called the mean absolute deviation, abbreviated naturally to MAD.

$$\text{MAD} = \frac{1}{n} \Sigma \, [x_i - M]$$

So for our fifteen-point data set, we get MAD from:

$$\frac{0.42 + 0.32 + 0.22 + 0.22 + 0.22 + 0.12 + 0.12 + 0.12 + 0.12 + 0.12 + 0.12 + 0.18 + 0.28 + 0.38 + 1.28}{15}$$

$$= 0.28$$

It is somewhat frustrating to find that MAD is seldom used. Statisticians prefer to eliminate the problem of positives and negatives by squaring the $x_i - M$ value. (Remember that when you square a negative, it becomes a positive—one of the little conventional joys that make mathematicians dewy-eyed.) So now we get the mean square deviation (MSD):

$$\text{MSD} = \frac{1}{n} \Sigma \, [x_i - M]^2$$

With the next bit, you are going to have to trust me without explanation. Suffice it to say that the MSD is converted into something called the standard deviation (s) by dividing not by the total number of measurements (n) but rather by one fewer ($n - 1$), and then taking the square root of the value that we so painstakingly squared!

$$s = \sqrt{\frac{1}{n-1} \; \Sigma \; [x_i - M]^2}$$

Once more we turn to our data set and get that s is the square root of:

$$\frac{\begin{matrix} 0.176 + 0.102 + 0.048 + 0.048 + 0.048 + 0.014 + 0.014 + 0.014 \\ + 0.014 + 0.014 + 0.014 + 0.032 + 0.078 + 0.144 + 1.64 \end{matrix}}{14}$$

$$= \sqrt{0.171}$$
$$= 0.41$$

s^2 is called the variance. Incidentally, sigma (σ) is often substituted for s to represent standard deviation. Life is nothing if not complex.

So, we have defined standard deviation and variance, now it is time to put these concepts to work in assessing the range of values about a mean.

Normal Distribution
It might be argued that normality is the exception rather than the....oh, we're heading in a circle here! It is customary for statisticians to talk of normal distributions, as depicted in Figure 3.3. It really is an illustrative way of describing probability, i.e., the likelihood of a value being a certain distance from the mean. Thus there is a 68.26 percent chance of a value being within one standard deviation of the mean, 95.44 percent probability of it being within two standard deviations and a 99.73 percent chance of it being within three standard deviations.

Figure 3.3 ━━
Normal Distributions as Measured by Standard Deviations

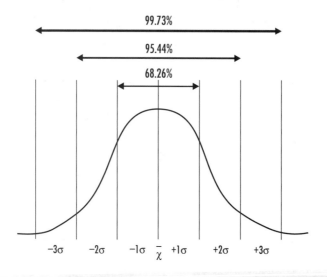

Much of what we encounter in our lives will be true to this "normal" pattern. Let's use the weight of the population to illustrate. If the actual weight of the entire male population of the United States was quantified, it would be expected to be plottable with a shape of the type seen in Figure 3.3. If, however, only a relatively limited and atypical sample was studied—let's say football players who enjoy their beefsteaks—the pattern would be skewed. Jockeys would also give a different subset.

Of course, to generate such a smooth curve demands measurements on an infinite number of samples. In practice, only a limited number of measurements is taken, so that the curve needs to be generated on the basis of the data available. As we will encounter repeatedly in the chapters that follow, the key is to select samples that are representative of the material we are trying to analyze, samples that encompass the breadth of data for any given parameter.

A random variable (x) becomes standardized when it has been adjusted so as to have a mean of zero and a standard deviation of one. The resultant value, z, is given by $(x-\mu)/\sigma$, where μ is the true mean. (μ is the "theoretical" mean, as opposed to M, which is the observed mean derived from actual observations.) z is called the standard normal variable.

Table 3.1 ▬▬▬▬▬▬▬▬▬▬▬▬▬▬▬▬▬▬▬▬▬▬▬▬▬▬▬▬▬▬
Normal Distribution Table

$z=\dfrac{(x-\mu)}{\sigma}$	A(z)	$z=\dfrac{(x-\mu)}{\sigma}$	A(z)
0.0	0.5000		
0.1	0.4602	2.1	0.0179
0.2	0.4207	2.2	0.0139
0.3	0.3821	2.3	0.0107
0.4	0.3446	2.4	0.0082
0.5	0.3085	2.5	0.0062
0.6	0.2743	2.6	0.0047
0.7	0.2420	2.7	0.0035
0.8	0.2119	2.8	0.0026
0.9	0.1841	2.9	0.0019
1.0	0.1587	3.0	0.00135
1.1	0.1357	3.1	0.00097
1.2	0.1151	3.2	0.00069
1.3	0.0968	3.3	0.00048
1.4	0.0808	3.4	0.00034
1.5	0.0668	3.5	0.00023
1.6	0.0548	3.6	0.00016
1.7	0.0446	3.7	0.00011
1.8	0.0359	3.8	0.00007
1.9	0.0287	3.9	0.00005
2.0	0.0228	4.0	0.00003

A (z) is the area under the standardized normal plot. Notice that the uppermost value is 0.50. This is when z = 0, i.e. when z is actually the true mean (center point, or peak, of the plot). Obviously for this value half of the plot (0.5 – or 50%) is to the left of this value and half to the right.

These values allow us to calculate the extent to which samples we are interested in are likely to be outside certain analytical limits. Tables with values that allow one to work out normal probabilities of this are available. (See Table 3.1.)

All this is a tad complex. I reckon an example is the best way to illustrate matters.

Let's say we were targeting a pH in our beer of 4.1 ± 0.1. (That is, the lower limit is 4.0 and the upper limit is 4.2.) Analysis over time has shown that the measured pH shows a normal distribution about a mean of 4.15, with a standard deviation of 0.05. How do we calculate what proportion of beer batches are likely to be outside the specified limits?

For the lower limit $x = 4.0$ and therefore:

$$z = \frac{(4.0 - 4.15)}{0.05} = -3.0$$

The minus sign indicates that we are looking at data below the mean. By reference to Table 3.1, we see that the probability level is 0.00135. In other words there is a 0.135 percent chance of a value being below this value.

For the upper level:

$$z = \frac{(4.2 - 4.15)}{0.05} = 1.0$$

Again referring to the table, we find that the probability level of a sample being too high in pH is 0.1587.

The total probability of a sample being outside acceptable limits of pH is 0.00135 + 0.1587 = 0.16005, with obviously the chances being greater for the pH to be too high rather than too low.

Let's take the issue further by taking a look at Figure 3.4. Two curves are shown, the broader one being classic normal distribution that would be obtained for an infinite number of

Figure 3.4

Variance versus Standard Error

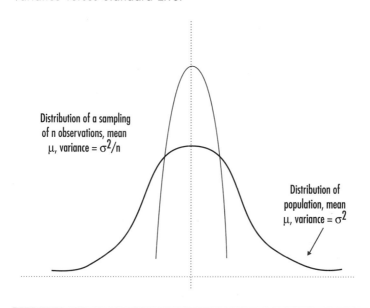

Distribution of a sampling
of n observations, mean
μ, variance = σ^2/n

Distribution of
population, mean
μ, variance = σ^2

samples. Its peak is μ. The variance (as we've just seen) is σ^2. Peeking at the narrower, taller curve, we see the picture obtained when measurements are on a finite number of samples (n). The mean should be the same (μ) but now the variance is described by σ^2/n. The n value reflects the fact that we are dealing with a limited, measurable number of samples. One more thing to grasp (or take on good faith!) is that the standard deviation of sample averages (otherwise known as the standard error) is given by $\sigma/\sqrt{n}$. (Compare this with the mean standard deviation referred to earlier.)

Enough of all this jargon—let's take a look at an example of the value of all of this in a practical situation.

We are producing bottles of beer that should each contain 330 mL. The legal requirement is that this must be the average net volume for a randomly selected sample of 50 containers. In other words, if 50 bottles are sampled, on average they must contain 330 mL. Let's say that the standard deviation of the filling process is 5 mL. At what volume would we need to set the filler to have a 0.95 probability that the minimum volume requirement is fulfilled?

The standardized normal deviate ($z_{0.05}$ to signify the 95% probability level—i.e., a 5% likelihood—$A(z)$ in Table 3.1—of a value being outside the target range) is given by:

$$\frac{(M - \mu)}{(\sigma/\sqrt{n})}$$

If you look at table 3.1, you will see that 0.05, falling between 0.0446 and 0.0548, gives a z value between 1.6 and 1.7. That value is actually 1.645—more detailed versions of the table allow us to come up with that number.

Inserting the values we now have into this formula we get:

$$\mu = 1.645 \times (5/\sqrt{50}) + 330$$

Now we calculate that the mean value (μ) needs to be set at 331 mL. Thus we would have to set the filler at this level to have confidence that we were (on average) filling at a sufficient level. For individual packages, the standard deviation is 5 mL.

Going back to the type of calculation we did in the pH example, the proportion of bottles that will be below the declared volume is obtained from:

$$z = \frac{(330 - 335.41)}{5.0} = -1.082$$

Reading from a more detailed version of table 3.1 I can tell you that this equates to a value of around 0.14. In other words, despite the risk of an average of 50 samples containing less than 330 mL of beer being only 0.05, in fact 14 percent of the individual packages will contain less than this volume. Remember that we are considering average contents, not minimum contents.

All of this has been a discussion of normal distributions. Without going into detail, I would just say that all distributions are not necessarily normal and for diverse reasons may be skewed. (See Figure 3.2.) It's all too complex to consider here.

Process Capability

A fundamental question arises: just how controlled can a process such as the various stages of malting and brewing be? Nothing that happens in this chaotic world of ours is immune from variation and some degree of inconsistency. Folks who talk about "Statistical Process Control" call it noise. There are two types of this noise:

1. Internal noise, examples of which include fluctuations within a batch of raw materials (malt, hops, yeast, water, etc.) and wear and tear in machinery.
2. External noise, such as different operators and differences between sources of raw materials (e.g., harvest year).

These variations are pretty much unavoidable. However, if the process generates a product that has a stable mean value, it is said to be in control. There may be a lot of variation about the mean, but only within clearly understood limits. Sometimes, though, a process deviates to a greater extent than this, leading to a shift in the mean or an increase in the variability. The process is then out of control. Process capability studies are devoted to understanding the extent of the variation that exists.

To assess the variability of a process we need to take samples that are representative of the range of values that can be expected. The more the merrier (as for anything where statistics are concerned) but 50 samples, taken over a period long enough to reflect time-dependent variation, seems invariably to be a good number. Making whatever measurement we need, (let's say it's dimethyl sulfide levels in a lager beer) we declare the range of results observed (R) and calculate the process mean (μ) and the standard deviation (σ). The data can be plotted, too, using curves of the type with which we have already become familiar. Figure 3.5 illustrates processes that display various degrees of precision. Remember that if we consider beer as the product of a single process, but evaluate its precision using different parameters measured on the beer, we might expect to see the breadth of plots as depicted on this diagram. Thus when measuring alcohol, we anticipate a very narrow plot (high precision). This is because we can control final beer strength very tightly. By contrast, a parameter such as dimethyl sulfide may show considerable

Figure 3.5

Curves Showing Different Degrees of Precision

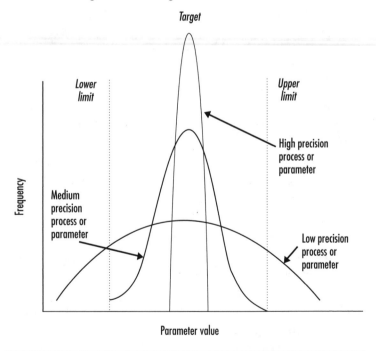

variation, in which case we would say that we had a medium precision process (or worse). A high precision process is one in which the variation in values is very small when compared against the specification limits. As we will see in the chapters that follow, the major parameters by which a brewer assesses beer acceptability (measures of strength, carbon dioxide and oxygen content, vicinal diketone level, color) are under pretty tight control. Measures such as certain flavor volatiles tend to be less well regulated. (Dimethyl sulfide is simply ignored by some brewers, for instance.)

If we wish to quantify the process capability, then it is given by:

$$\frac{\text{Upper limit of measurement} - \text{lower limit of measurement}}{6\sigma}$$

The denominator is 6x standard deviation because (as we saw in Figure 3.3) this range pretty much describes the full range of values in a standard distribution.

It is only through observation and measurement that judgments can be made about what range of values can be tolerated in any process. Some parameters are measurable to a high degree of precision and can be regulated very tightly, percentage alcohol for example. In these instances, the control charts are tall and narrow. Other measures are substantially less precise and the likely spread in a product such as beer may be considerable. Foam stability measurements tend to be one example.

Control Charts

Charting data make it easier to spot when a process is moving out of control. When properly done, these charts allow the operator to determine whether any changes are within or without normal variation. If a change is observed in a parameter but it was entirely within normal fluctuation, it would be foolhardy to tinker

Figure 3.6
Control Diagram with Warning and Action Lines

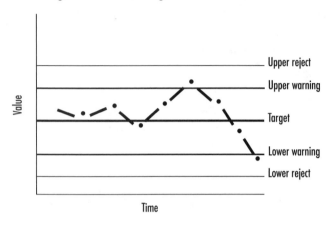

with things. The frequency of data plotting should reflect outputs. For some parameters (i.e., alcohol content), every batch of beer will be measured so the data is there to be plotted. By contrast, flavor compounds (i.e., the esters) may only be measured on occasional batches. This will define their plotting frequency. It is only those parameters whose value will lead to a go/no go decision that should be measured on every batch.

The simplest type of control diagram to use is to plot data within a framework of action (reject) and warning lines. (See Figure 3.6.) Convention has it that these reject lines should be positioned at three standard errors ($3\sigma/\sqrt{n}$) above and below the mean. Again we need to remember that this explains 99.73 percent of the variation in data for a normal distribution. It is often the case that two other "tram lines" within the action lines are used (warning lines). These are set at two standard errors ($2\sigma/\sqrt{n}$). My personal experience is to call the action and warning lines the Reject Quality Limit (RQL) and Acceptable Quality Limit (AQL) respectively.

Another type of plot frequently used is the CUSUM (cumulative sum) plot. This is analogous to keeping score in golf, where you gauge performance on the basis of how much above or below par you are.

Let me first give you the algebraic definition of CUSUM and then make sense of it with an example:

$$CUSUM = \Sigma_i (x_i - T)$$

where x_i is the ith measurement of a quantity and T is the target value (what the measurement ideally should be or specification). So the CUSUM is the summation of all the deviations from normality. To take an example, let's say that our target is 6, and these are ten successive measurements made:

5, 8, 6, 7, 6, 9, 4, 5, 5, 7

Then the CUSUM is calculated as shown in Table 3.2.

Table 3.2 ━━━━━━━━━━━━━━━━━━━━━━━━━━━
Calculation of CUSUM

Observation number (i)	x_i	x_i-T	CUSUM $\Sigma(x_i$-T)
1	5	-1	-1
2	8	2	1
3	6	0	1
4	7	1	2
5	6	0	2
6	9	3	5
7	4	-2	3
8	5	-1	2
9	5	-1	1
10	7	1	2
x = measurement T = target			

An example of a CUSUM plot is shown in Figure 3.7. These plots are valuable for highlighting where substantial changes are occurring in a parameter. When the slope of these plots trends upwards then this indicates that the average is somewhat above the target value. When the slope trends downwards this indicates that the average is somewhat below the target value. The more pronounced the slope, the more the value deviates from target.

Perhaps the ultimate control chart for measured variables is that which plots both the mean and the variability (average and range). Enough sampling needs to be done for there to be confidence in the mean and range values, but not to the extent that "real time" information is not being obtained. This approach lends itself to stages in a process where there are many samples available. For example, in a brewery with one lauter tun but a carousel bottle filler, *in extremis* we might have 20 or 30 measurements for the bottler for every one on the fermenter. There are very few measurements to capture what is happening in a fermenter, such as daily testing of specific gravity. By contrast, the frequency of checks on a filler is likely to be much greater.

Figure 3.7
CUSUM Plot

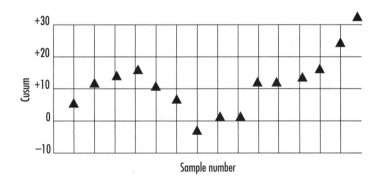

Table 3.3 ━━━━━━━━━━━━━━━━━━━━━━━━━━━━━━━━
CUSUM Calculations for Figure 3.7

The measurement is for foam stability using the Rudin method. Each week a sample of the beer is measured, the target is 100 seconds. The data generated over 15 weeks is as follows:

Observation number (i)	x_i	x_i-T	CUSUM $\Sigma(x_i$-T)
1	105	5	5
2	106	6	11
3	103	3	14
4	101	1	15
5	97	-3	12
6	94	-6	6
7	92	-8	-2
8	103	3	1
9	101	1	2
10	110	10	12
11	100	0	12
12	101	1	13
13	104	4	17
14	107	7	24
15	108	8	32

Applying This to the Brewing Industry

The average brewery generates oodles of data: numbers describing raw materials; process information; product analyses; performance indices (i.e., yield and efficiency). I have always adhered to one overriding credo: unless you intend doing something with a number (or are obliged to collect it—such as for legal reasons), don't collect it, don't save it, don't worry about it. Our lives contain enough clutter without worrying about extraneous information that we are simply going to look at and file away.

The second truism is that the most valuable data to collect is that which increases the chances of other numbers being within the prescribed target range. If by measuring something upstream and acting on it, values downstream come into specification, then the former measurement is the more critical. An example would be the control of mashing temperature. By worrying about and controlling that temperature, we can have confidence that the wort will have the desired level of fermentability, and our yeast will produce alcohol and flavor compounds in the appropriate balance. The mashing temperature is, of course, not the only critical measurement— others will include assays on the yeast itself. The key issue is that there are a relatively few *critical control points* that need to be quantified and regulated for our process to be standardized. Some people talk about process control and product control. It will be apparent that good process control should greatly increase the chances of product control.

I spoke earlier about control charts and how they are used to illustrate how well our processes are doing within the framework of acceptable variation. I defined the limits of variation as typically being up to three standard errors. (See Figure 3.6.) If the values are within these tramlines then the process is said to be under control. It is important to emphasize, however, that a process classified as under control by these criteria may not necessarily produce a good product. Consider for instance a flavor compound such as dimethyl sulfide. Some brewers like it in their beer, some don't. Let's say that our company is striving towards a level of 50 ppb. It is extremely unlikely that levels in beer could be controlled to better than ±10 ppb. (Indeed it probably can't be measured with that degree of accuracy—as we shall consider later.) I can assure you that a beer containing 40 ppb DMS and one containing 60 ppb DMS are likely to be told apart by many tasters. And lest we forget, lots of other flavor compounds will be fluctuating in a similar way (most of which haven't even been identified, let alone measured). So although a process yielding a lager consistently within the tramlines of 40 and 60 ppb for a target of 50 ppb would be said to be under control, it does not necessarily mean that all batches of the product would be indistinguishable for the parameter in question. In other words, regulation of DMS to within ± 2 ppb (which is presently unrealistic and probably always will be using current brewing paradigms) would represent a more controlled process. If, however, the control was within this range 90 percent of the time but the rest of the time (and at random), values for DMS were way out, say by 50 ppb, then the process would be out of control.

Establishing Controlled Systems

Figure 3.8 illustrates the principles of achieving control over our processes. Fundamentally, the loop involves measuring a key parameter and adjusting the process to counter any drift from the desired level of that parameter. An example would be temperature control in a fermenter. If a temperature rise is detected (a natural event due to metabolic energy produced by yeast), cooling is applied to counter heat accumulation. Naturally, the brewer will have identified in advance what the temperature regime should be in order for the desired product to be reached.

The most sensitive measurement/adjustment loops are those made during the process with a close link between measurement and adjustment. Thus, for instance, it would be strange to wait for lower than specification bitterness unit values in packaged beer before hopping rate was adjusted. We encountered a classic example earlier with the control of yeast pitching.

One word I find myself using repeatedly in my teaching is compromise. Every drop of beer in the world is produced as a compromise. Maltsters tolerate losses through embryo growth

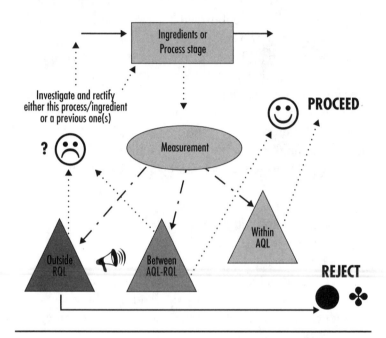

Figure 3.8

Achieving Control Over Process: A Flowchart

in order that decent modification is achieved in germination. Wort boiling is a balance between achieving sterilization, isomerization, volatilization, control over flavor compounds such as dimethyl sulfide and concentration on the one hand, and avoiding thermal stress and off-flavor production on the other. And so on. Therefore our control loops must recognize this. If the feedback from one measurement leads to a process adjustment that will invariably throw another parameter out of specification, this needs to be accounted for in whatever loop is established.

Chapter Four ▄▄▄▄▄▄▄▄▄▄▄▄▄▄▄▄▄▄▄▄▄▄▄▄▄▄
Standard Methods of Analysis

The brewing industry has a long pedigree of establishing standard methods of analysis. One of the problems is that national and international preferences and prejudices are at play. It means that worldwide there are several sets of methods with many overlaps and significant differences, with origins lying in the brewing practices of separate nations. Thus the standard methods of the Institute of Brewing (now called the Institute and Guild of Brewing, IGB) were originally developed in England for the analysis of ale-type beers and so, for example, use small-scale infusion mashes at constant temperature. The methods of the European Brewery Convention (EBC) are based on the production of lager-style products: their mashes (Congress mashes) have a rising temperature regime. There are major efforts to meld the two different compendia.

As my yardstick in this book I am taking the Methods of Analysis of the American Society of Brewing Chemists (ASBC). (See Bibliography.) These procedures probably owe more to the EBC than the IGB but they form the analytical reference point for all North American brewers, whether producing lagers or ales. They have been shortened into a volume (also listed in the Bibliography) that ought to be particularly valuable to craft brewers.

Whether we are talking IGB, EBC, or ASBC, the usual approach to the establishment of standard methods is to use a committee. An old boss of mine was solidly of the opinion that the most effective committee to get things done consisted of one person. It has been said that a camel is a horse designed by a committee. However in the cause of establishing standard methods of analysis, a committee is rather useful. In particular it will comprise folks in different laboratories who can test the method out.

The sequence of events goes something like this: First, a method has to be nominated. Some bright young spark, for example, has developed a procedure that she believes is the ultimate for measuring fiber in beer. She has used it extensively in her research and made all manner of claims when using it. She says it really "kicks ass." The question is: Do other people get the same results when using it? So she sends it to the society committee—or they go in search of it and ask her to send it, because they are interested in getting hold of such a procedure.

If she's anything like the rest of us boffin-types, her method will not necessarily be written in a way that is fully comprehensible and readily followed. I learned this lesson years ago when a method I suggested was totally redone and put into a form that, at the time, I said was written for idiots. That was a tad cruel. Writing the methods down in a painful step-by-step way is an attempt to ensure that everybody performing the method follows exactly the same protocol. If there is any ambiguity then somebody somewhere will do things differently and get a different result.

Step two is for the method to be written out in a standard format. I used to joke that protocols start with "Come in to work at 8 a.m. ± 1 minute. Take off your coat. Put on your lab overall....." Of course, they don't—but that will give you a sense of their inherent regimentation.

Now the method needs to be tested out. Volunteers are sought from all those labs that have an interest in the analyte concerned. Many of the volunteer companies will be seated around the committee table. Each of them will be sent the standard method with a set of samples that have been produced in one location with the best aim of consistency (in this case beer for the fiber analysis).

Once analyzed, the results are returned to a central coordinator who crunches the data. In a moment I will talk about this aspect, and which values tell us whether the method is or is not going to be taken on-board as a standard procedure. Basically if a sufficient number of laboratories report good agreement for a range of samples, the method will pass. There may be one or two labs that have results way off, and this will be revealed. If everybody's data is scattered then it begs questions about the method rather than the laboratory.

Figure 4.1
Accuracy versus Precision

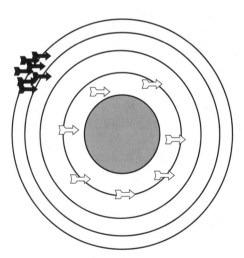

The archer using solid arrows is quite precise in hitting the same region of the target, but is away from the "correct" result. The one using outlined arrows is accurate insofar as on average the arrows circle the "mark" but with significant scatter between individual hits.

The Interpretation of Analytical Results

Think of archery. (See Figure 4.1.) Picture the target. Our first archer is firing arrows and groups them all together on the board but still the grouping is way off target. Clearly he is a very precise bowman, albeit one not capable of hitting the right value, namely the bullseye. Imagine another archer aiming for bullseye and landing a dozen arrows close in around the center with roughly similar spacing, though none of them hits the preferred mark. This archer is accurate insofar as she is closer to the true mark than the other one is but she is less precise. The ideal (and not one readily attained in the worlds of archery or brewing analysis) is to hit the bullseye every time. Essentially it would take forever to get to this perfection so the practical reality is that we have to get as good as we can in terms of accuracy (getting close to the truth) and precision (doing it consistently).

To dwell again in areas we addressed in the previous chapter, statisticians speak of variance when talking of the errors involved in any quality control test.

$$\text{Variance} = \frac{\text{the sum of (individual results} - \text{mean result)}^2}{\text{Number of results} - 1}$$

There may be several sources of this variance. One of them has to do with sampling. Imagine if we were analyzing a batch of barley, and the pile of grain was not homogeneous. If the pile wasn't mixed before taking samples for distribution to the different laboratories, each lab might get significantly different grains and in all good faith offer up different results. Throughout the book I talk about representative sampling and looking after samples to avoid such variance.

The other sources of variance (replication error and systematic error) have diverse origins: different batches of chemicals, unconscious deviations from the laid down method, contaminating species in glassware or water, atmospheric conditions, human imperfections and so on. The extent to which each of these matters depends on the method: the more robust the procedure, the less the scatter.

The best approach to testing for errors was developed by Youden. In this technique the various labs are sent a pair of samples that represent two different levels of the analyte under examination. For example it could be different levels of soluble fiber. Each lab would be asked to make their measurements on this parameter in the two beers—the values for beer one being the A series and those for beer two being B values. The collated data is then plotted as shown in Figure 4.2. The circle has a radius that is determined by multiplying the standard deviations of the replication error. (See Table 4.1.) Essentially there is a 95 percent probability of a result falling within the circle.

Figure 4.2 ━━━━━━━━━━━━━━━━━━━━━━━━━━━━━━━━━━━━━
A Youden Plot

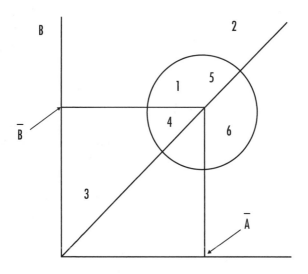

Each number represents the data plot for one laboratory. Lab 2 has a replication problem with B having a high value but not A. Lab 3 has a systematic error with low results for both beers.

Table 4.1 ━━

Using the Youden Method to Calculate Errors

	Laboratory 1	Laboratory 2	Laboratory n	Total	Average
Result A	A_1	A_2	A_n	ΣA	$\bar{A}$
Result B	B_1	B_2	B_n	ΣB	$\bar{B}$
A - B	X_1	X_2	X_n	ΣX	$\bar{X}$
A + B	Y_1	Y_2	Y_n	ΣY	$\bar{Y}$

Standard deviation of the total error = $\dfrac{\Sigma(Y-\bar{Y})^2}{2(n-1)}$

Standard deviation of the replication error = $\dfrac{\Sigma(X-\bar{X})^2}{2n-1}$

Two other values are important. The first of these is repeatability (r), which is an index of how consistently data can be generated by a single operator in a single location using a standard method. To use formal jargon: the difference between two single results found on identical test material by one operator using the same apparatus within the shortest feasible time interval will exceed r on average not more than once in 20 cases in the normal and correct operation of the method. Trust me when I tell you that it is given by $2 \times \sqrt{2} \times \sqrt{(s_r)}$, where s_r is the standard deviation for the procedure when assessed within a single laboratory. The second parameter is reproducibility (R), this being the maximum permitted differences between values reported by different labs using a standard method. This is the value obtained in collaborative trials between labs and is given by $2 \times \sqrt{2} \times \sqrt{(s_b^2 + s_r^2)}$, where s_b is the between lab standard deviation. In short, r is a measure of "within lab" error, R of "between lab error," and the lower r and R, the more reliable is the procedure.

Setting Specifications and Monitoring Performance

The information gleaned in this type of intercollaborative exercise helps the brewer achieve the goal of setting realistic and meaningful specifications for the product, the raw materials, and the process samples.

Let's take an example. From an intercollaborative exercise it has been shown that the measurement of ethanol by gas chromatography over the range 0.9 to 6 percent (v/v) has r of 0.061 and R of 0.136. (That's what was actually observed in a trial reported in the Journal of the Institute of Brewing 1993, 99, 381-384.) In other words, for a single operator within a laboratory the difference in results obtained on successive measurements of the same beer is going to be less than 0.061 nineteen times out of twenty, provided the method is used properly. For operators between laboratories the result will be within 0.136 of target with that same frequency. And so for this parameter if one lab gets a result of 5.00 percent and a second lab 5.05 percent, we can conclude that there is no significant difference between the results. Only if the difference is greater than the R value can we be confident that there is a difference in the result.

When setting specifications there are two key requirements:

a. knowledge of these r and R values, which indicate how reliable a method is

b. appreciating the true range over which a parameter can vary before a change is observed in quality.

Again, let me illustrate. Say the brewer has set a specification for the precursor of dimethyl sulfide (a key determinant of lager flavor as we will see) of 6 micrograms per gram of malt (μg/g), plus or minus 1 μg/g (6 ± 1 μg/g). This means that she expects the precursor level in the malt to fall within the range 5–7 μg/g. This has been decided after extensive trials which have shown that if the level falls between 5 and 7, it doesn't cause a perceptible change in DMS in the final beer. However, if the value is lower than 5 or higher than 7, too much or too little DMS is produced respectively. If the r value for DMS parameter is low (say 0.02), then to a first approximation any value falling outside the 5–7 range would clearly be established in the brewer's own lab. And if the R value was only about 0.04 then again the brewer could pick the maltster up on a value falling outside the range. But what if the method for measuring this precursor is lousy (and some would say it is)? Say that r is 1.0. Then the brewer's own lab could have very little confidence in reporting a value of, say, 4.5 as being outside specification. In reality it might just as easily be 5.0. Only if the lab analyzed the sample *ad nauseam* could a meaningful mean value be proffered. And say the R value was 1.0. In this case a brewer might get a value of 4.5 and start the verbals on the maltster, who would retaliate with,"No way, Jose, we find that this malt contains 5.0."

To avoid grief between maltster and brewer, brew house operative and cellarman, right on down the line, get methods you can trust and which mean something.

Chapter Five ▬▬▬▬▬▬▬▬▬▬▬▬▬▬▬▬▬▬▬▬

Barley

Although I confess to not having counted them individually, there are approximately 25 million kernels in every ton of barley. That's quite a lot. And the remarkable thing is that they're individually just as different as the folks in a sell-out at Yankee Stadium. Sure they're all recognizable as barley kernels, just as the baseball fans are all recognizably human (or at least very nearly so). Yet they all have their uniqueness, their subtleties of composition.

The Natural Variation Of Barley

There are several reasons why there are variations in any batch of barley. For starters, there may be some admixture of varieties, and these will behave just as differently as will different nationalities of people. Within a variety, though, there is plenty of scope for variation, including changes taking place within the barley as it is stored; changes that are due to hormonal fluctuations that determine amongst other things its ability to germinate. A single variety will analyze differently if it is grown in different locations with different soils and climates. In a single location the composition of a given variety will differ depending on the conditions it encounters each growing season, such as weather and whether it is a relatively wet or dry year. Even on a single ear the grain will differ. Those kernels lower down on the ear tend to be bigger. In a 6-row barley where there is less room for individual kernels than for 2-row barleys, there is considerable disparity between the corns depending on their location on the ear. Even across the starchy endosperm of each barley kernel there is a greater or lesser degree of variation in structure and composition.

If we are to make sense of any batch of barley, then, we have to get as representative a sample as possible of that grain. In practice this means that samples need to be taken from various places in a delivery of grain, whether shipment has been by railcar or truck. Samples will be taken by plunging a long hollow spear (called a trier) into the grain at various depths and lateral positions in the whole. The grain drops through holes along the length of the spear, and then a cover is slid over the holes. The trier is withdrawn, and the individual samples are mixed together evenly, and an overall analysis is taken. This will plainly give an average snapshot of what that batch of grain comprises. Only by analysis of the individual trier samples could there be some indication of the extent to which the batch varies—maltsters tend to refer to this as the extent of heterogeneity.

Analyzing Incoming Barley

Every self-respecting maltster will run a battery of checks on all shipments of barley as they arrive each season at the maltings. The results from these checks will be compared with the specification that has been agreed between maltster and merchant. If some specifications aren't met, the barley will be "returned to sender." For instance if the barley is of the wrong variety (in part or in its entirety), or dead, or infected, then the maltster will not accept it. There may be some parameters for which, if out of specification, a compromise may be reached with the supplier. For instance, if the nitrogen content is a little high, then a lower price might be negotiated by way of compensation.

The critical checks need to be made rapidly. Following the barley harvest there will be a continuous line of railcars or trucks loaded with barley arriving at maltings every day. These need to be checked in (or out) quickly so the checks to hand must be swift.

The first thing that will be assessed is the appearance of the grain. Is it a desirable amber yellow or are there distinct signs of infection? Greenness may be an indication of premature harvesting whereas dull grayness might be a symptom of weather damage. To what extent do foreign bodies litter the sample, from rocks through alien plant life to deceased rodents? Are there any signs of weevil infestation? Does the grain smell sweet? Are there signs of damage, for instance broken corns?

The truck will be tare weighed to ensure that the agreed weight of grain is present. Otherwise the three most important checks are for nitrogen content, moisture, and viability.

Nitrogen is a major component of protein, but not of starch. The protein content of grain is important as a source of the amino acids needed by yeast and also of foam backbone material. However it is the starch that the brewer is particularly interested in, because that is what he is going to break down in the brew house to produce the fermentable sugars that the yeast will be turning into alcohol. Ergo the more starch the better. For a given size of barley kernel, the more nitrogen (and therefore protein) is present, the less room there is for starch. Which is why maltsters seek barley of relatively low nitrogen content.

The measurement of nitrogen is not an especially rapid procedure, as we shall see a little later when discussing the individual tests that are applied. Maltsters, therefore, are grateful for three big letters: NIR. That's Near Infrared Reflectance spectroscopy to the cognoscenti. (See Figure 5.1.) In a nutshell, chemical substances will absorb light in the near infrared region (between the visible part of the spectrum and the infrared region) with characteristic spectra. By measuring the NIR spectrum of a compound or mixture of compounds you can get information about what is present and in what quantity. For 20 years now maltsters have been applying NIR spectroscopy to

Figure 5.1

The Principles of Spectroscopy

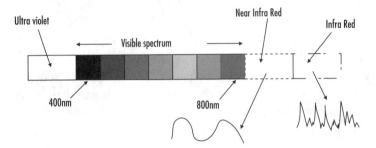

The electromagnetic spectrum can be divided into the familiar visible spectrum and into ultra violet light (UV, at lower wavelengths) and infra red (IR, higher wavelengths). Compounds absorb light characteristically at different wavelengths.

measure the level of protein and moisture in barley samples. It used to be that the barley was ground up to a powder before analysis, but remarkably the measurements can be made on whole corns. The sample is placed in the machine, and the optics and programs therein flash up the nitrogen and water information in seconds. The machine will first have been calibrated by putting in samples of known nitrogen and moisture content so the relationship between spectral data and the absolute level of these components is established.

The barley has to be alive or it won't convert into malt. The quickest way to assess this viability is using a dye called tetrazolium. The barley kernels will be sliced in half lengthwise and flooded with the dye. If the embryo is alive then it will decolorize the stain, but if the grain is dead then the embryo stains crimson.

If the barley passes the visual inspection and satisfies the maltster for its nitrogen, moisture and viability, it will be accepted into the maltings and transferred to silo storage. There is time then for a somewhat more leisurely analytical check of the grain.

The Comprehensive Analysis Of Barley

The laboratory tests that are applied to barley to assess its suitability for malting (and therefore brewing) can be divided into those assessing the overall physiology and generic properties of the grain and those chemical tests that quantify specific components within the grain.

Physiological And Anatomical Tests

The maltster needs to have confidence that the barley in her silo is of the desired variety. The first way to do this is by skilled visual inspection and checking certain anatomical characteristics of the grain. For varieties grown in North America, the relevant bible is, *Know Your Malting Barley Varieties*, a pamphlet published by the American Malting Barley Association. (AMBA; 740 N. Plankinton Avenue, Suite 830, Milwaukee, WI 53203) It is also available on their web site at www.ambainc.org. Experts, armed with the descriptions and photographs in this publication as well as with a magnifying glass, can pick out the variety and determine whether a sample contains more than one variety. They'll be looking for characteristics such as wrinkling of the hull (husk), whether there is any slight twisting (which indicates 6-row types), whether there is a long or a short rachilla, and if the hairs on it are short or long. (Let's not worry about precisely what a rachilla is—but whether a barley has got one of a certain length and whether it is hairy or not seems to be somewhat important for some.) The analyst may pearl the barley to expose the aleurone layer and see whether it is the customary white or if it is blue, which is the case for a very few cultivars.

Every year I expose my students to the challenge of telling varieties apart in this way and each time they screw up. I can't be critical because I would fail the test myself. (Luckily, I get to look at the crib sheet sent by the kindly maltsters who donate the grain.) It really isn't an easy task to complete. We are grateful, therefore, for the advent of protein and DNA fingerprinting techniques. In just the same way that a forensic scientist would use such approaches to solve crimes, so can a barley laboratory apply them to identify barley varieties and check for the sin of admixture. The proteins or DNA are extracted from grain and separated by electrophoresis (the fractionation of chemical species on the basis of their differential migration when exposed to an electric field). The patterns obtained are a characteristic of the barley variety concerned.

The next question is: how big are the kernels in this barley sample? Fundamentally, bigger kernels are preferred because they contain pro rata less husk and therefore more starch. However the natural heterogeneity of barley means that any sample contains corns that differ to a greater or lesser extent in size. Because thinner kernels take up water more rapidly than bigger ones, it is desirable to keep the grain of different sizes apart for steeping, in the interests of achieving consistent malt for brewing. Most maltsters in fact will reject samples containing too high a percentage of "thins". Taking the weight of a thousand kernels assesses the overall size of grain, a value cunningly called Thousand Kernel Weight (TKW). Actually it is a pain to count out such a large number of corns, so the approach is to weigh out 15g and then to count the number in that sample. The weight of any damaged kernels that have survived screening processes is subtracted, and the value is corrected to account for the moisture content of the grain. The assessment of the distribution of kernels of various sizes (Assortment) is determined by sieving the grain. For six-row barleys the slot widths are 2.38mm (6/64 inch) and 1.98mm (5/64 inch). Those kernels held back by the first sieve are said to be plump, and those evading the second sieve are called thin. Two-row barleys tend to be bigger than 6-row ones (there's more room for them on the ear), and so for them the second sieve is of 2.18mm (5.5/64 inch). A more detailed sieving may be used, using more sieves with a greater range of slot widths.

Apart from size, another property that impacts the ability of barley to take up water in steeping is the texture of the starchy endosperm. If it is fine, white, and crumbly (mealy), water will be taken up readily, and the endosperm will tend to be readily degraded subsequently. However if the endosperm is hard, gray, and glassy (steely) , water uptake and ensuing digestion will be difficult. One test for this property is to pearl the barley and visually inspect the grain, splitting it into piles of kernels that are a) $\frac{3}{4}$ or more steely, b) $\frac{1}{4}$ to $\frac{3}{4}$ steely, c) $\frac{1}{4}$ or less steely. The categories are called glassy, half glassy, and mellow respectively. Shining light through the pearled barley can facilitate the differentiation. The glassy kernels tend to reflect the light and appear light. The mealy (mellow) corns appear dark.

The next battery of checks is designed to assess the germination characteristics of the grain. Is the grain alive? Is it ready to be used?

The definitive test of germinability of grain is Germinative Capacity. One hundred barley kernels are placed in a beaker and soaked in a solution of 0.75 percent hydrogen peroxide. (Nobody is absolutely sure of the role that the peroxide plays, but it is suspected that it overcomes any dormancy in the barley. Remarkable stuff, though, peroxide: it also bleaches hair and can put rockets into space.) The number of barley kernels which chit (i.e., in which the rootlets start to emerge from the tip of the grain) is counted after 72 hours (or earlier if information is desired about how vigorous the germination is). Most maltsters will set a specification of greater than 96 percent for this parameter, in other words no more than 4 of the kernels should fail to chit over the three days.

Even if a batch of barley is alive, it may not necessarily germinate. It is said to be dormant. This is a natural condition, a precaution built into the grain to prevent it from germinating on the ear. Lest the maltster or brewer ever forget it, barley exists to make more barley, not inherently to make beer! It is

in the barley plant's interest for its seed to germinate only at
the appropriate time, and that is when it has escaped from
"home" (the parent ear) and is in a nice cozy place ready to
set down roots. Barley as it arrives in the maltings will display
different degrees of dormancy, depending on the variety and
the environmental conditions that it has been exposed to. In
any case, dormant barley will not make good malt for the
simple reason that it won't germinate. Thus the maltster
needs tests to advise him when a barley has become freed
from dormancy. The extent of dormancy is assessed in the
Germinative Energy test. A filter paper is soaked in 4 ml of
water and placed in a covered dish. One hundred kernels are
distributed on the paper. The number sprouting is counted
after one, two, and three days. The target is greater than 96
percent germinating after three days. Clearly the number has
to be taken alongside the Germinative Capacity test. Thus if
the latter test gives a value of 97 percent and so too does the
Germinative Energy test, we can conclude that the batch is 97
percent living corns, none of which are dormant, and three
percent dead grain. If the Germinative Capacity test read 100
percent but the Germinative Energy test 49 percent, we can
say that all of the grain is alive but roughly half of the sample
is dormant. By counting the number germinating at one-day
intervals, the maltster can get a measure of the vigor with
which the grain is germinating. This test will be run regularly
throughout the storage time of the barley. As barley is stored
it becomes more vigorous in its germination capabilities. By
monitoring this in respect of the rate of chitting over one, two,
and three days of the Germinative Energy test, the maltster
can get clues about how to perform the steeping and
germination operations. They will test these out on a small
scale (micromaltings) before translating them into modified
commercial regimes targeted on achieving the desired
specification in the finished malt.

A third germination test assesses water sensitivity. If grain encounters too much water then it may germinate less well. It's almost as if the grain is swamped. The extent of this water sensitivity is assessed by repeating the Germinative Energy test but with 8 ml of water instead of 4 ml. The existence of the phenomenon explains why most malting operations involve interrupted steeping regimes, where periods of water contact are interspersed with air rests. The more water sensitive barleys as judged by this test will receive a greater number of shorter steeps.

The last test in this battery is the assessment of pre-germination. Occasionally barley kernels do germinate prematurely while still on the ear in the field. This is a major problem because that grain will die before coming into the maltings. There are no entirely reliable methods for assessing pre-germination, but most people will use a stain that registers whether enzymes have already been developed in the grain. Barley kernels are embedded in a clay or plastic bed, and after this has solidified, the grains are sanded down to half their depth using a commercial sander (I kid you not). The grain is then flooded with a solution of the dyestuff fluorescein dibutyrate, and after 10 minutes the number of kernels showing yellow fluorescence adjacent to the embryo is taken as an indication of pre-germination. They fluoresce because in germinating they have produced an enzyme that breaks down the dyestuff.

Chemical Tests
The standard chemical tests on barley are made after it has been milled to a fine powder. The maltster is concerned that the water content of the grain (moisture) shouldn't be too high for at least two reasons: a) she wants to buy grain not water b) too high a water content means too great a risk of the grain germinating prematurely. It is also essential to know the moisture content so other analyses can be corrected to a dry weight basis and meaningful comparisons can be made between different samples.

Moisture is simply assessed by drying the milled grain. The sample is weighed before and after keeping it for an hour in an oven at 266 °F (130 °C). The difference in weight represents the water driven off.

The nitrogen level of the grain was for many years assessed by a method developed by Kjeldahl of the Carlsberg Laboratory in 1883. In this procedure, the sample of barley is cooked with acid to release ammonia from its protein and then the ammonia assessed by titration. In this way a value for the total level of nitrogen (N) is generated. As many folk prefer to talk in terms of protein, that level is computed by multiplying the nitrogen value by 6.25 (on the basis that the average protein is 100/6.25 = 16% nitrogen). In times when people are increasingly concerned about health and safety, there has been a move away from the Kjeldahl method and its unpleasant sounding ingredients. It has been superceded by a procedure based on the total combustion of a sample in pure oxygen, a method devised by Jean Baptiste André Dumas. The nitrogen released is assessed using a thermal conductivity detector. The values generated tend to be somewhat higher than those obtained with the Kjeldahl method. Accordingly, when the latter method was first introduced, the naïve started to criticize their suppliers for trying to pass off higher nitrogen grain. In reality the grain hadn't changed at all, only the method had. There is a salutary lesson here: for many of the measurements made in the malting and brewing industries (and in most other walks of life) there are few absolute values. Rather there are numbers that mean something to those who understand both the procedure used to get those numbers and the meaning of them in the context of the business. It's akin to currency. The value of a dollar can go up and down but the thing it buys isn't changed. As long as we agree on the value involved and all parties understand the level at which it is set, there is no problem. Or picture it in terms of language: if we decided to move away from English to, say, German, foam is still foam, only now it's *Schaum*. As long as you know that foam is *Schaum* we can get along famously.

Maltsters seldom, if ever, directly measure the level of starch in barley. In fact the only specific class of carbohydrate that they might measure is β-glucan, the major component of the cell walls in the starchy endosperm. Low levels suggest (but don't guarantee) that endosperm breakdown will be easy and that there may be fewer troubles in brewing. Many believe that mealy barleys tend to contain less β-glucan, such that the assessment of the degree of glassiness is probably a good indicator of potential glucan problems. It is probably informative of all aspects of maltability and brewability. There is an unwritten rule of thumb in all matters analytical that the more methods there are for measuring a single component, the poorer they all are. Why else would you have so many methods if just one was reliable enough to do the job? So we find that there are several methods for measuring β-glucan. Some people swear by Calcofluor, a material that binds to β-glucan and fluoresces. So by measuring the fluorescence of extracts of barley to which Calcofluor is added you can assess how much glucan is in those extracts. (Incidentally Calcofluor is the sort of material that has been used in those whiter-than-white washing detergents, the sort that make your cotton shirts fluoresce when you go disco dancing—which I don't. Cotton has a structure similar to barley β-glucan.) Another method for assessing β-glucan is to hydrolyze the glucan using specific enzymes and measure the glucose produced.

Much more important as a routine measurement for the maltster is to assess the safety quotient on the grain. The initial inspection of the grain should have given a strong indication about its hygiene status. In particular the maltster is concerned about fungal infection of the grain, especially infection with Fusarium. This organism is particularly prevalent in regions that are relatively damp, so it had not been a special concern for North

American growers. Until a few years ago, that is, when there were extensive problems with the organism. This was a direct result of changed agricultural practices. The burning of residual straw in fields after harvest was outlawed in favor of plowing it back into the land. The fire had served to destroy fungal spores but now these were surviving to contaminate the next year's crop. Fusarium is a problem for at least two reasons. The first is that it produces toxins, notably deoxynivalenol (DON). Secondly it produces small proteins that cause beer to gush. These proteins pass through the malting and brewing processes and, if they get into beer, there is no alternative but to destroy it. Hence the absolute requirement to avoid the use of grain that is infected with Fusarium. There is no good direct test for the gushing factor in barley but DON can be measured in extracts of barley using gas chromatography. If DON is detected, it is an indicator of gushing problems but the presence of DON itself would be sufficient reason to reject that barley, because of its health connotations.

Exercises

1. Three barleys have been analyzed with the following results:

Parameter	Barley A	Barley B	Barley C
TKW (g)	40	43	41
Germinative Energy (%)			
24h	62	43	51
48h	95	77	70
72h	95	88	83
Germinative Capacity (%)	96	89	98
Kernels germinated after 72h in 8ml water (%)	91	82	45
Moisture (%)	12	19	11
Nitrogen (%, as is)	1.7	2.1	1.9
Mealy kernels (%)	89	75	37
Steely kernels (%)	11	25	63

a. Which barley has the lowest viability?
b. Which barley is most dormant?
c. Which barley is most water-sensitive?
d. Which barley would you predict to have the lowest β-glucan content?
e. Which barley displays the most vigorous growth?
f. Which barley has not been dried?
g. The viable kernels of which barley might be expected to give the highest levels of enzyme after malting?
h. Which barley has the biggest kernels?
i. What alternative procedure could have been used to come to this conclusion?
j. Which barley would you say had the least digestible endosperm?

2. How much protein is present in a barley registering as 1.72 percent nitrogen (dry weight basis)?

3. For the three barleys listed in the table in Question 1, what is the average moisture content of each kernel in terms of milligrams?

4. The anatomical characteristics of some UK barley varieties have been tabulated:

Variety	Halcyon	Pipkin	Prisma	Fanfare
Aleurone color	Blue	White	White	White
Rachilla	Short/medium length with long hairs	Medium/long with long hairs	Medium/long with long hairs	Medium with long hairs
Grain size	Medium	Small	Large	Medium-large
Special characteristics			Husk loosely attached	Hairs on the ventral furrow

Identify the following mixtures of barleys and determine which variety predominates:

    a. Sample A has a TKW of 46 g and on inspection there appears to be a significant degree of surface damage. There is a distinct coloration within cross-sections of some of the grains. The rachillas are quite lengthy for the most part.

    b. Sample B has a TKW of 42 g. Perusal of the crease indicates some degree of hairiness. The rachillas are generally quite lengthy.

5. Critically discuss the barley samples for which data is reported here:

	Barley A	**Barley B**	**Barley C**
Total Nitrogen (%, as is)	1.7	2.1	1.4
Germinative Capacity (%)	97	99	90
Germinative Energy (%)	5	98	90
Assortment (%)			
Retained at 2.38mm	46	10	52
Retained at 1.98mm	91	82	99
Thousand Kernel Weight (g)	43	39	45
Glassy (%)	15	61	5
Half glassy (%)	30	20	6
Mellow (%)	55	19	79
Staining with fluorescein dibutyrate (%)	1	0	11
Total moisture (%)	12	11	13
β-glucan (%, as is)	4.1	5.9	3.7
DON (mg/kg)	<0.1	0.9	<0.1

6. Two samples of barley have been distributed by a central laboratory to four satellite laboratories for analysis. The data generated is tabulated here:

	Central Lab		Lab A		Lab B		Lab C		Lab D	
Nitrogen (%)	1.61	1.95	1.62	1.93	1.72	2.03	1.62	1.94	1.63	1.95
β-Glucan (%)	4.1	6.2	3.3	4.7	5.0	5.5	4.3	6.0	2.1	6.4
DON (mg/kg)	0.8	0.1	0.9	n.d.	0.8	n.d	0.9	0.1	0.9	n.d.
Moisture (%)	12.3	13.0	12.5	12.6	13.0	12.9	12.6	12.6	12.9	12.6

n.d. = non-detectable

Critically discuss the data.

Chapter Six

Malt and Adjuncts

I t's hugely unfortunate, but perhaps the biggest unwritten rule in the world of brewing is "blame the maltster." The maltster has for too long been the whipping boy (or girl), condemned for all manner of sins that are visited upon brewers by way of poor yields, slow-to-collect wort, sticking fermentations, and lousy flavor. I won't say that poor malt won't cause problems because it will. I must say, though, the conversion of barley to beer is a shared responsibility of maltster and brewer. Just as many (if not more) evils are perpetrated in the brew house and cellar as can be foisted on the shoulders of the honest maltster.

We are dependent on identifying good robust methods of malt analysis that help the maltster control her process, tell the brewer of a malt's suitability for use and serve as the basis of contracts and purchase at the maltster-brewer interface.

Sampling Malt

Just as for barley, representative samples of the malt must be taken. This applies to malt coming into the brewery as well as for malt leaving storage and headed for milling. Silos are around that are capable of holding as much as 30,000 tons of grain (and each ton of malt will occupy about 1.8 cubic meters). To fill such a silo with malt would take a substantial number of deliveries (a truckload may comprise 20 tons), each of which must adhere to specification. But within specification means on average, and there may be some differences from batch to batch. Even leaving aside the likelihood of admixture in the silo as grain settles, it should be evident that as the silo is emptied (with perhaps more malt being put on top) there will be a greater or lesser degree of variation in the grain headed for weighing and milling. The greater the differences from batch to batch and within batches entering the silo, the greater the variation of malt headed for processing.

Once again we encounter the concept of heterogeneity. Brewers would rather talk of homogeneity. What they are seeking is homogeneous malt in all its manifestations so they can predict as well as possible how it will behave in the brewery.

We have already seen in chapter 5 how there are many sources of heterogeneity in barley. Every kernel is different to a greater-or-lesser degree. It follows, therefore, that these differences will be passed on to the malt, unless in some miraculous way the malting process equalized the differences. It doesn't, rather it exaggerates them. For instance, there are differences in the time each kernel encounters the various process stages, such as whether it is soon onto the kiln at the end of the germination phase or one of the last to be transferred to the heating process.

All of which means that we need representative samples of malt entering the brewery and, ideally, of malt headed to the mill. As yet the latter is impractical. Not only do we need representative samples, but we also need a strong indication of the range of values within a given sample. In other words, we need to know how homogeneous it is.

Let's take an example (and it serves us well as an illustration of the phenomenon of data spread that we encountered in chapter 3). Imagine that we are measuring β-glucan. (See Figure 6.1.) We have two malts and in both of them we register somewhere between 0.5 and 1.0 percent β-glucan. Does this mean we can rejoice in the prospect of no sticky problems?

Figure 6.1 ━━━━━━━━━━━━━━━━━━━━━━━━━━━━━━━━━━━

The Importance of Distribution Differences

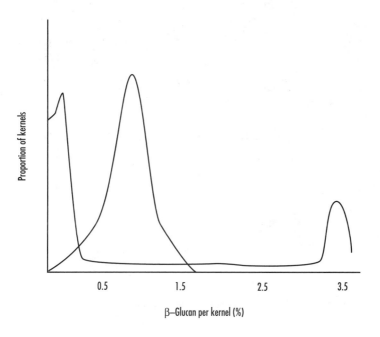

These curves show the distribution of β-glucan values in two separate samples of malt that each have identical average (overall) β-glucan content.

Nope. For it may be that in the first sample we have got pretty much all the kernels modified to the same extent, with a degree of residual cell wall material at the distal end of the kernels. In the second, though, we may have the vast majority of the kernels extremely well-modified with virtually no surviving cell wall material. However, a small proportion of the kernels, perhaps a tenth of them, have hardly been modified at all, perchance because the starting barley contained excessive levels of dead or dormant grain. These unmodified barleys will have high glucan and will be hard and difficult to mill. In the case of the first malt, we can rig up the mill to deal with that degree of modification because it is uniform. For the second malt, however, this setting will tend to pulverize the overly modified grain. This will produce fines that will likely clog up the wort separation stage. However, very coarse particles will be generated from the grossly under-modified kernels, which will leach troublesome β-glucan into the wort but will otherwise be resistant to degradation and leave unconverted starch and potential extract in the spent grains. So we need tests that will detect heterogeneity and, before that, sampling regimes that will present a fair picture of the material that we are dealing with.

Sampling of deliveries of malt can be performed using long triers (6 feet or even 12 feet) for railcars or trucks or short triers (3 feet) for bags. For grain in transit, for example en route to a silo, trickle samples can be taken.

Physical Tests
Some of the same tests that are made on barley will be made on malt. These measurements include Thousand Kernel Weight (comparison of this with that of the original barley indicates how much weight has been lost in the malting process), presence of foreign matter, and kernel size distribution.

The texture of the starchy endosperm will also be assessed. Traditionally a skilled maltster can do this with tools no more sophisticated than his teeth and the tender rubbing of green malt between thumb and forefinger. Somewhat more sophisticated for the finished malt is the assessment of mealiness in kernels that have been sliced in half longitudinally. One hundred kernels will be inspected in this way and scored as mealy, half-glassy, or glassy, in decreasing order of their modification. This is not an especially easy judgment to make, even with the aid of a needle to prod and poke the endosperm.

To aid the differentiation we can turn again to Calcofluor. The 100 kernels are solidified in a plastic matrix and sanded down to half size, before flooding with Calcofluor. If the plate is then inspected under a fluorescent light, the well-modified kernels will appear dark, whereas any undegraded β-glucan-rich cell walls will be revealed as fluorescence. The method is very dependent on good sampling and on the orientation of the kernels—whether they are back- or belly-side-up makes a difference. The glucan can be measured in worts, too.

Another popular technique for assessing modification involves use of a friabilimeter. In this method the kernels are pressed between a rubber roller and a rotating sieve drum. Properly modified malt will be broken into particles that will pass the sieve. This amount by weight is quoted as percentage friability, a value that correlates with overall extent of modification. A further refinement of the test is to take the residual coarser fraction and pass it through a second stage using a 2.2 mm wide sieve. This will retain any grossly under-modified kernels (anything bigger than a half kernel) and the percentage retention by this sieve is an indication of severe under-modification.

Another valuable technique for assessing modification is by measuring the length of the acrospire (the shoot) that grows under the husk. A sample of the malt is boiled for half an hour, in which time the husk is loosened. The length of the germ is counted in 100 kernels and classified according to 0-$\frac{1}{4}$, $\frac{1}{4}$-$\frac{1}{2}$, $\frac{1}{2}$-$\frac{3}{4}$ and $\frac{3}{4}$-1, in terms of the length of the acrospire in relation to the length of the kernel. The longer the acrospire, the greater the modification. Note is also made of overgrowth, involving those kernels where the acrospire is longer than the kernel itself.

Chemical Analysis

Just as for barley, the maltster is interested in knowing the moisture and nitrogen content of the malt. The methods involved are essentially the same. The moisture content reflects the amount of kilning that the malt has received. It must not be too high or the malt is increasingly susceptible to spoilage. Malt with too high a water content is said to be slack.

The major analysis performed on malt, however, involves mashing it on a small scale. The most controlled way to do this is by using a mashing bath, which allows for the side-by-side extraction of a large number of samples (for example, about 50 g of each) under carefully controlled conditions of temperature and agitation. The malt is milled, typically using a disc mill, and then mixed in the beakers with a defined quantity of water to achieve the desired solid-liquid ratio. Ideally this mimics the mash thicknesses used commercially but frequently is one that is relatively thin (the ASBC standard mash uses 200 ml water to hydrate the 50 g of malt). Various temperature regimes are possible in such baths so that the mash can be performed isothermally (at a constant temperature) or using a ramped temperature regime designed to match some of the more complex mashing programs employed in production brewing. Note will be taken of the odor of the mash—it should be aromatic and hopefully not musty, green, stale or otherwise unpleasant. Once the mash has reached conversion temperature (i.e., 158 °F (70 °C)) the time for the starch reaction with iodine to disappear will be noted. As iodine stains starch blue, small quantities of mash are placed on a white tile and mixed with iodine solution. Conversion is reached when a blue coloration no longer appears. After the allotted period of mashing some extra water is mixed in (cold water in order to return the temperature to ambient relatively quickly), and the contents of the mashing beaker are poured into filter funnels lined with filter paper. The time taken to collect a certain volume is measured, or the volume collected in a set period is noted. Either way, this is a rough (some would say too rough) way of predicting how worts from that malt will lauter in the brewery. Note should be taken of the clarity of the wort, whether it is bright, slightly hazy, or hazy.

The wort obtained is analyzed for various components. (Bear in mind that the range of parameters I am going to unveil now may also be measured on worts derived in commercial wort production systems, in order to ensure that the wort is fit for use. We'll come back to this in chapter 9.) At the top of the list is the specific gravity, assessed traditionally using a pycnometer or hydrometer but increasingly more often by digital density meter. (See chapter 11.)

This value for specific gravity is in proportion to dissolved materials in the wort. In particular, the more sugar dissolved in the wort by the action of enzymes principally in mashing, the greater will be the specific gravity. The specific gravity number is converted into a value for extract.

In the case of the ASBC mash the calculation is performed as follows:

The total weight of wort is given by $[(M + 800) \times 100] / (100-P)$

where M is the percent moisture content of the malt (which must be accounted for if we wish to compare malts on a dry weight basis), and P is grams of extract in 100 grams of wort (otherwise known as °Plato).

$$\text{Extract (as is) (E; \%)} = (P/100) \times \text{total weight of wort}$$
$$= [(P \times (M+800) \times 100]/[100 \times (100-P)]$$
$$= [P(M+800)] / (100-P)$$

$$\text{Extract (dry basis) (\%)} = 100E / (100 - M)$$

The specific gravity (SG) is converted into °Plato by tables but the rule of thumb is to subtract 1, multiply by 1000, and divide by 4. To be rigorously accurate the brewer should measure °Plato directly using a hydrometer so-calibrated, or else use the equation:

$$°P = \{258- [205 (SG-1)]\}(SG-1)$$

To me, dividing by four is easier!

For example a specific gravity of 1.0440 is converted to Plato in this way

$$1.0440 - 1.0000 = 0.0440$$
$$0.0440 \times 1000 = 44$$
$$44/4 = 11$$

Thus a wort of specific gravity 1.0440 is close enough 11 °Plato—although the deviation between these methods becomes noticeable above about 13 °P (1.052).

Extract is a critical parameter, for it is on the basis of extract that barley varieties are selected (only the ones yielding high extract after malting and mashing will be selected), by which malt is traded and by which the strength of the wort stream heading off into the nether regions of the brewery will be gauged.

Yet it is not only the specific gravity (and hence extract) that is assessed on the wort generated in this small-scale mash. Various other parameters are quantified, including measures that give some indication of the extent to which the malt proteins have been degraded during malting and mashing. First there may be an assessment of how much nitrogen is dissolved in the wort. Just as for barley, this was formerly ascertained by the Kjeldahl procedure, but nowadays is likely to be performed by the Dumas method.

This value is divided by the figure for the total nitrogen in the malt and expressed as a percentage in a ratio called either the Soluble Nitrogen Ratio or the Kolbach Index. The difference is that the latter is the value when the mashing is done using the Congress mash of the European Brewery Convention, which is a relatively complex protocol with several temperature stages designed to reflect the decoction mashing approaches of European brewers. In either case the figure is an indication of how much of the protein in the barley has been solubilized during malting and mashing. The protein level in wort (and thence beer) is important because proteins are on the one hand backbone materials in foam while on the other hand, bad news for beer haze. However, there is no simple relationship between the soluble nitrogen ratio and either of these aspects of beer quality, so I for one have always doubted its real value.

Probably of more use is Free Amino Nitrogen (FAN). The proteins of barley need to be converted to amino acids before the yeast can use them, so the level of amino acids present in wort is of particular significance. FAN is measured by adding the reagent ninhydrin to a portion of the wort. This material generates an intense violet color when it reacts with amino acids.

No reactions are necessary to measure several other substantive components of the wort. pH is assessed using a pH meter, which must be properly calibrated using standard solutions of a pH close to that of the liquid being assessed. For wort, this means a buffer of pH probably around 5.5, whereas for beers, the buffer should be of the order of 4 to 4.5. Remember that pH changes with temperature, and so wort at 149 °F (65 °C) will have a pH about 0.35 lower than that measured at 68 °F (20 °C). As for all measurements, it is important that the person trying to make sense of them knows exactly how the numbers were generated. I will discuss pH in rather more detail when talking about water in chapter 7.

The viscosity of wort is an index of whether there are significant levels of β-glucan present and may be used by the brewer as a forewarning of potential wort separation and beer filtration difficulties. Viscosity is measured typically using a glass viscometer, which assesses the rate at which liquid flows through a capillary tube, relative to a standard of water. Temperature is very important—the higher the temperature, the lower the viscosity.

Viscosity = (flow time of wort at 68 °F (20 °C) / flow time of water at 68 °F (20 °C)) x wort specific gravity x 1.002

where 1.002 is the viscosity of water at this temperature. The units of viscosity here are centipoise (cP). Obviously the higher the specific gravity of the wort the greater the viscosity is likely to be, so to compare worts this needs to be taken account of.

Another parameter measured on the wort with a minimum of additional preparation is color. This is assessed by measuring the amount of light that wort absorbs at a wavelength of 430 nm. I'll describe in chapter 11 why this is somewhat over-simplistic but for pale lagers and lager worts it is probably an adequate technique. Because wort, no matter how visually bright, will pretty much always contain some particles that will scatter light and masquerade as color by increasing absorbency at 430 nm, it must be filtered before the spectrophotometry. A little diatomaceous earth is added, the mix is swirled and filtered through paper.

Some maltsters and brewers require detailed analysis of some of the chemical species present in malt or derived from malt during mashing. The wort from the small-scale mash is used to measure some of these. Above all, there is interest in the levels of fermentable and non-fermentable carbohydrates that are present. The most informative way of measuring these is by using chromatography, either gas chromatography or, more frequently, high performance liquid chromatography. Each sugar has a characteristic retention time under a given set of chromatographic conditions. By measuring the size of its peak in comparison to defined amounts of known standards, an accurate assessment of the amount of each sugar present can be derived. Some maltsters and brewers don't have the luxury of sophisticated chromatography set-ups. If they're interested in fermentability, they assess it by measuring the extent to which added yeast reduces the specific gravity in a small-scale fermentation. The specific gravity is measured before and after the fermentation and, by difference, an estimate of the proportion of extract that is fermentable may be obtained. There are a number of shortcomings with the procedure: the results are dependent on the yeast used. Other factors, such as nutrients in the wort also impact the result. Furthermore the procedure is relatively slow, taking at least two days whereas chromatographic measurement of sugars takes hours at most.

The metal ions present in malt and wort can be assessed using atomic absorption spectrophotometry. Of particular interest is zinc, which many brewers add as a nutrient for yeast.

Let's get back to measurements made on the malt itself as opposed to those made on the wort from the small-scale mash. I have discussed the significance of extract, which of course is primarily derived from starch by enzymic hydrolysis during mashing. Apart from containing a high level of starch (ergo low nitrogen, see chapter 5), malt must contain sufficient quantities of the enzymes that are needed to degrade the starch. If it doesn't then the malt will have to be rejected or the enzymes will need to be supplemented with those of exogenous origins. The enzymes responsible for the degradation of starch are the amylases, primarily α-amylase and β-amylase. The first of these converts starch into dextrins, and the other chops up starch and dextrins to maltose, for the most part. Both are needed for efficient starch degradation but it is the former that is largely responsible for converting starch into a form that no longer reacts with iodine and β-amylase that is responsible for developing most of the fermentability. Together they are assessed as diastatic power. The standard procedure for gauging this parameter is to incubate an extract of malt with a standard starch solution. The reducing sugars produced are measured and are in proportion to the total amount of the two chief starch-degrading enzymes present. An assessment of α-amylase alone is obtained by the dextrinizing units method. Here a substrate is used that has been treated already by β-amylase, and an excess quantity of that enzyme is added. The β-limit dextrin substrate is mixed with the malt extract, and the rate of loss of iodine staining potential is assessed. The more quickly the color is lost, the more α-amylase is present.

The only other enzyme likely to be measured in malt is endo-β-glucanase. This is the chief enzyme responsible for breaking down those pesky β-glucans in the cell walls of malt. Ideally they are dealt with in the germination phase of malting but some may survive and it may be necessary for more enzyme degradation to occur in mashing. Endo-β-glucanase is extremely sensitive to heat, and so kilning has to be performed carefully to avoid excessive loss of the enzyme. In the brew house, mashing may need to be started at a relatively low temperature (i.e., 104 to 122 °F (40 to 50 °C)) to allow the enzyme to survive (it is rapidly destroyed at conversion temperatures). β-glucanase can be measured in several ways. The traditional technique is to monitor the drop in viscosity of β-glucan solutions caused by extracts of malt. The second is to measure the release of reducing sugars from β-glucan by malt extracts. A third is to assess the release of dye from a specially prepared dyed-version of β-glucan. Finally, β-glucan can be reacted with Congo Red and then incorporated into a gel in a dish. Small holes are bored in the gel, and the malt extract added. It diffuses into the gel, breaks down the β-glucan, which, in turn, loses the red coloration from the Congo Red. A zone of clearing is produced, the diameter of which is in proportion to the amount of enzyme.

Just as the maltster is concerned with the safety of the incoming barley, so is the brewer eager to know that the malt is safe and wholesome. DON will be specified just as for barley. Equally important in the malt specification is an assessment of nitrosamines. These potentially carcinogenic substances (proven in the rat, not in the human) develop during malt kilning. Precursors in the embryo of malt react with nitrogen oxides in the gases on the kiln to produce nitrosamines. Modern malting practices have virtually eliminated nitrosamines as a problem but this certainly does not mean that the brewer has lost interest in measuring them. The malt is extracted, and the nitrosamines fractionated by gas chromatography followed by detection using thermal energy analyzers.

Finally let's turn to flavor. Essentially the only parameter measured on malt that has anything directly to do with flavor is S-methyl methionine (SMM). When this material (which develops in the embryo during germination) is heated it breaks down to dimethyl sulfide (DMS), a key component of the aroma of many lager beers. Whenever a heating stage occurs (malt kilning and wort boiling primarily) SMM breaks down to DMS. Those brewers wanting to eliminate DMS as a flavor constituent are likely to be interested in substantial heat treatment, but are also interested in less SMM being developed during germination. Equally, those brewers desiring a finite level of DMS also need a reliable assessment of SMM. This can come from methods based on high performance liquid chromatography, but much more common is a technique whereby SMM is extracted in water and then heated in a sealed vessel under alkaline conditions. This releases DMS which is measured by gas chromatography, detection either being with a flame photometric detector or a chemiluminescence detector. SMM in wort is measured in the same way. (Another precursor of DMS, called dimethyl sulfoxide, does not as yet form a part of most brewers' specifications for malt.)

Adjuncts

Many of the same principles that apply to the measurement of malt pertain to the evaluation of adjuncts. Adjuncts of course may be either solid (raw or micronized grain, grits, flakes, flour) or liquid (syrups or sugars). Meaningful and representative sampling is a given. Physical evaluation of the adjunct is equally important: does the adjunct look and smell/taste as it should, with no off colors or moldy, musty, or rancid aromas? Are there any curious looking foreign bodies in the sample?

Moisture is assessed on a solid adjunct by an oven drying method just as for barley and malt. Protein is measured by comparable methods. The extract available from a solid adjunct is slightly more problematic, insofar as for the most part adjuncts are devoid of enzymes. These need to be supplied, either from a proportion of malt or by the addition of commercial amylases. The small-scale mash is conducted much as for the evaluation of malt but with differences depending on the adjunct. Because the starch in some adjuncts (i.e., rice, corn, and sorghum) has a very high gelatinization temperature, the cereal is boiled prior to cooling and addition of the bulk of the malt to enable mashing. Adjuncts that are pre-gelatinized (i.e., flaked or torrefied cereals) are mashed-in directly with the malt. Flakes, of course, are not milled. The calculation of the extract that originates in the cereal involves subtraction of the extract afforded by the malt itself.

Parameters measured on solid adjuncts that tend not to be assessed on malt are oil and ash. Oil can be quite a substantial component of adjuncts such as rice, presenting a risk to foam and flavor. It is assessed by extracting it from milled cereal using petroleum ether, evaporating this solvent and drying and weighing the residue.

Ash is essentially the material left behind after you have burnt everything else off (at 1022 °F (550 °C)!)—just like the embers from your campfire. If you incinerate a material such as a cereal then the carbohydrates, proteins, lipids, nucleic acids, and assorted other organic molecules burn up to produce carbon dioxide, nitrogen oxides, and water. The inorganic materials such as metals that are not converted into evaporating volatile oxides are assessed by weighing what is left behind.

In some ways liquid adjuncts are easier to analyze than solid ones. Thus extract determination does not involve any mashing, merely dilution of a given weight of the adjunct and assessment of its specific gravity. In turn the percentage moisture content of liquid adjuncts is determined by simple subtraction from 100 of the percent extract ($^\circ$Plato) of the adjunct.

Liquid adjuncts can be produced to different degrees of fermentability so estimation of fermentable extract is frequently significant. As for malt, it is assessed by monitoring the drop in extract caused by yeast in a small-scale fermentation or from a direct assessment of the sugars by a technique such as high performance liquid chromatography.

Ash and protein in liquid adjuncts is determined as for solid adjuncts.

As many liquid adjuncts are prepared commercially by using controlled acid hydrolysis of starch (although, increasingly, specific amylolytic enzymes are used), the assessment of the pH and acidity of the adjunct can be important. pH is assessed using a pH probe dipped into a 10 percent solution of the adjunct (50 grams of syrup made up to 500 ml with water—this is the same solution that is used to assess extract and the other salient parameters in the liquid extract). This solution is titrated with caustic to assess the degree of acidity in the sample.

Exercises

1. Three malts (D, E and F) have been analyzed with the results shown on page 82.
   a. Which malt appears to have the greatest modification (on balance)?
   b. List three parameters that have led you to this conclusion.
   c. Which malt might have been produced from barley of low Germinative Energy?
   d. Which malt has probably received the most kilning?
   e. Which malt is most likely to lead to filtration problems downstream?
   f. Which malt contains most starch-degrading enzyme activity?
   g. Which malt (under appropriate conditions) would be best able to deal with a barley-based adjunct? (continued on page 83)

Parameter	Malt D	Malt E	Malt F
Moisture (%)	4.0	2.7	5.3
Extract (fine, % as is)	75.9	76.3	71.7
Extract (coarse, % as is)	74.7	73.3	67.4
Filtration volume (ml collected in 30 min.)	127	105	97
Acrospire length (%)			
0 to ¼	5	2	7
¼ to ½	10	13	19
½ to ¾	42	52	61
¾ to 1	42	30	13
>1	1	3	0
Nitrogen (%, as is)	1.79	1.76	2.2
Wort-soluble Nitrogen (%, as is)	0.077	0.078	0.066
Free amino nitrogen (mg/100 g malt, dry)	153	164	125
Diastatic Power (°, as is)	155	142	131
α-Amylase (DU)	45.1	44.4	42.0
β-Glucanase (IRVU)	550	450	450
Color (°)	1.80	1.95	1.78
NDMA (ppb)	1.7	<1.0	<1.0
Friability (%)	89	86	74
Friabilimeter-crushed malt retained on 2.2mm screen (%)	1	1	6
Viscosity (cP)	1.6	1.65	1.95
SMM (mg/g)	10.2	3.2	5.4
DON (mg/kg)	1.7	<0.1	<0.1

The term "as is" means this is the value produced in the assay — it has not been corrected for the moisture content of the malt. For example for malt D the extract value (coarse) quoted on a dry weight basis would be 74.7 x 100/96 = 77.8%. For malt E it would be 73.3 x 100/97.3 = 75.3%.

h. Which malt is most likely to lead to a high DMS level in the finished beer?

i. Which malt is most likely to lead to a gushing problem in beer?

j. Which malt is the least likely to have been kilned using indirect heating?

2. Here is a partial analysis of two malts (Malt A and Malt B):
   a. Which sample has less well-modified cell walls?
   b. Which sample has more extensively modified protein?

Parameter	Malt A	Malt B
Moisture (%)	2.0	3.0
β-Glucan (as is)	0.5	0.5
Soluble nitrogen ratio	44	44

3. Here are some measurements made on adjuncts. Identify each adjunct from the list below.

	A	B	C	D	E	F	G	H	I
Extract (% dry)	78	78			79	79.5			
Moisture (%)	4.2	6.5	1.7	1.9	1.95	3.85	12	5	9
Color (°EBC)	180	35	1310	1120	65	18			
Viscosity (cP)						1.7	1.4	2.3	1.4
Oil (%)							0.73		0.4

Black malt; maize grits; crystal malt; torrefied barley; amber malt; Cara Pils malt; flaked rice; Munich malt; chocolate malt

Water

An old boss of mine, possessed of a vast intellect but little appreciation of the beauty of brewing, once said that beer is slightly contaminated water. Naturally I took issue with this description, whilst accepting the undeniable, that most beers are at least 90 percent water. The ones that aren't (those containing, should we say, warming levels of alcohol) are for laying down or laying down after.

And so the brewer cannot escape the need to take good care of the water supply, remembering too that more water is needed for brewing beer than ends up in the product. A well run, environmentally conscious brewery might use five times more water than finishes in the bottle, can, or keg. At the other extreme, as much as twenty times more water is used.

The quality criteria for water in the brewery might vary depending on whether that water is or is not going to end up in the beer. Fundamentally, if the water reaches those parts of the plant through which the process stream will pass, it ought to be of the highest chemical and microbiological quality. This, then, will include all water used to clean vessels and pipes. Only water that will not access the product or its production stream should be held to a lesser standard (for the most part microbiological). This would include water for cooling towers, raising steam in boilers, hosing floors, and putting out fires. Even here, though, chemical and microbiological concerns should be manifest. Water in cooling towers might harbor *Legionella*. Hard water will scale up pipes. And so on.

I'm no Moses: I tend not to part waters. It's invariably best to treat all waters the same—save for that water which needs to be deaerated (e.g., for diluting high gravity beers) or deionized (e.g., for dissolving isomerized hop extracts). Much more detail about the chemistry of water will be found in Appendix 3.

The basic premise for water used in the brewery is that it should not harm humans and neither should it screw up the plumbing. It should fulfill all legal requirements both chemically and microbiologically as well as satisfy the brewer's standards for clarity, lack of color, taste, and smell. Every self-respecting brewer should taste the water supply and that in storage as a matter of daily routine. In fact, that's just about the only routine check I would advocate on water, other than pH (as an overall index to show that nothing totally screwy is happening). For the fact is that of all the raw materials used to make beer, water is perhaps best regulated and least likely to be variable.

In the United States water must satisfy the National Primary Drinking Water Regulations established by the Environmental Protection Agency. These are summarized in Table 7.1. Additionally there are National Secondary Drinking Water Regulations. (See Table 7.2.)

The onus of ensuring the highest quality of water supply is for the most part shifted from the brewer to the supplier because most brewers do not use their own wells. This does not absolve brewers from ensuring that they don't do anything to detract from the quality of the water they have purchased—such as microbe pick ups from less-than-pristine tanks or iron from filters. Accordingly it is appropriate for periodic confirmatory analytical checks to be run on water at various stages in the brewery, which for many brewers will involve sending samples to a water analyst. The chances are that pretty good consistency will be observed between measurements.

Water analysis is a reasonably specialized issue. The microbiological methods employed are comparable to those that I describe in chapter 11. However, the organisms under scrutiny, such as *Legionella*, coliforms, *Giardia*, and *Cryptosporidium*, are different from those which concern the brewer as wort or beer spoilers.

The inorganic constituents of water, cations (positively charged), and anions (negatively charged) are also assessed by the type of methods used for analyzing these materials in beer. (See chapter 11.) Simpler, titration-based procedures of greater antiquity may still be used in some quarters—procedures such as determination of total hardness by titrating with the chelating agent EDTA, which binds calcium and magnesium. Once the ions are fully bound, surplus EDTA is detected as a green color by special indicator materials. Similarly total alkalinity can be determined using titration with the indicators phenolphthalein and methyl orange. Organic contaminants demand a diversity of specialized chromatographic set-ups.

Turbidity of water is assessed by nephelometry. The materials responsible for flavor taints are for the most part chlorinated byproducts of disinfection or industrial chemicals and apart from being detected by taste will be located by chromatography.

Once again, let me say that the parameter most readily measured by the brewer is pH, whether by pH meter or even indicator paper. Of course, the ultimate criterion for the acceptability of water for brewing is its taste. The biggest risk of taint is from chlorinated materials, hence the need for robust filtration of water through activated carbon, the check for efficiency of which is taste and appearance. Batch to batch differences in ionic composition of incoming water are unlikely to be dramatic. It is unlikely that they will be of a magnitude likely to cause exaggerated differences in process performance or product quality. The brewer should take comfort in a cordial relationship with the water supplier and place the onus on them. Depending on the nature of the water in a given locality, it may be necessary to make adjustments to parameters such as pH or hardness. (See Appendix 3.) But it will invariably be the case that the range of variation in water composition in a given location will not be so great as to warrant any brew-to-brew adjustment in the treatment.

Table 7.1 ━━
Extract from the National Primary Drinking Water Regulations

Component	Maximum contaminant level goal	Maximum contaminant level (mg/L unless stated)	Potential health effects	Sources of contaminant
Crypto-sporidium or *Giardia*	zero	99 – 99.9% removal/ inactivation	Diarrhea, vomiting, cramps	Fecal waste
Legionella	zero	Deemed to be controlled if Giardia is defeated	Legionnaire's Disease	Multiplies in water heating systems
Coliforms (including *Escherichia coli*)	zero	No more than 5% samples positive within a month	Indicator of presence of other potentially harmful bacteria	Coliforms naturally present in the environment; *E. coli* comes from fecal waste
Turbidity	n/a	<1 nephelometric turbidity unit.	General indicator of contamination, including by microbes	Soil runoff
Bromate	zero	0.01	Risk of cancer	Byproduct of disinfection
Chlorine	4	4	Eye/nose irritation; stomach discomfort	Additive to control microbes

Component	Maximum contaminant level goal	Maximum contaminant level (mg/L unless stated)	Potential health effects	Sources of contaminant
Chlorine dioxide	0.8	0.8	Anemia; nervous system effects	Additive to control microbes
Haloacetic acids (e.g., trichloracetic)		0.06	Risk of cancer	Byproduct of disinfection
Trihalo-methanes		0.08	Liver, kidney or central nervous system ills, risk of cancer	Byproduct of disinfection
Arsenic		0.05	Skin damage, circulation problems, risk of cancer	Erosion of natural deposits; runoff from glass and electronics production wastes
Asbestos	7 million fibers per liter	7 million fibers per liter	Benign intestinal polyps	Decay of asbestos cement in water mains; erosion of natural deposits
Copper	1.3	1.3	Gastro-intestinal distress, liver or kidney damage	Corrosion of household plumbing systems; erosion of natural deposits
Fluoride	4	4	Bone disease	Additive to promote strong teeth; erosion of natural deposits

Component	Maximum contaminant level goal	Maximum contaminant level (mg/L unless stated)	Potential health effects	Sources of contaminant
Lead	zero	0.015	Kidney problems; high blood pressure	Corrosion of household plumbing systems; erosion of natural deposits
Nitrate	10	10	Blue Baby Syndrome	Runoff from fertilizer use, leaching from septic tanks, sewage, erosion of natural deposits
Nitrite	1	1	Blue Baby Syndrome	Runoff from fertilizer use, leaching from septic tanks, sewage, erosion of natural deposits
Selenium	0.05	0.05	Hair or fingernail loss, circulatory problems, numbness in fingers and toes	Discharge from petroleum refineries, erosion of natural deposits, discharge from mines
Benzene	zero	0.005	Anemia; decrease in blood platelets; risk of cancer	Discharge from factories; leaching from gas storage tanks and landfills

Component	Maximum contaminant level goal	Maximum contaminant level (mg/L unless stated)	Potential health effects	Sources of contaminant
Carbon Tetrachloride	zero	0.005	Liver problems; risk of cancer	Discharge from chemical plants and other industrial activities
Dinoseb	0.007	0.007	Reproductive difficulties	Runoff from herbicide use
Dioxin	zero	0.00000003	Reproductive difficulties, risk of cancer	Emissions from waste incineration and other combustion; discharge from chemical factories
Alpha particles	Zero (as of 12/8/03)	15 picoCuries per liter	Risk of cancer	Erosion of natural deposits
Beta particles and photon emitters	Zero (as of 12/8/03)	4 millirems per year	Risk of cancer	Decay of natural and man-made deposits

The full table can be found at www.epa.gov/safewater/mcl.html. There you will find all the other items not listed above, a total of 63 other line items, the majority being a range of industrial and herbicidal chemicals.

Table 7.2
National Secondary Drinking Water Regulations*

Contaminant	Secondary standard
Aluminum	0.05-0.2 mg/L
Chloride	250 mg/L
Color	15 color units
Copper	1 mg/L
Corrosivity	Non-corrosive
Fluoride	2 mg/L

Contaminant	Secondary standard
Foaming agents	0.5 mg/L
Iron	0.3 mg/L
Manganese	0.05 mg/L
Odor	3 threshold odor number
pH	6.5 – 8.5
Silver	0.1 mg/L
Sulfate	250 mg/L
Total dissolved solids	500 mg/L

*These are non-enforceable guidelines regulating contaminants that may cause cosmetic effects (i.e., skin or tooth discoloration) or aesthetic effects (taste, odor, color). States may choose to adopt them as enforceable standards.

Chapter Eight

Hops

My wife threw it away after one sniff.

I speak of the after-shave produced from hops given to me by one hop processor eager to find an alternative use for the precious plant. As I write neither they nor anybody else has succeeded in finding any outlet for hops other than brewing. For which, in one way, the brewer should be thankful, for if the hops folk succeed in their quest then the worm may turn, and brewers may no longer have so powerful a position in buying "for peanuts" a material that makes a disproportionate contribution to beer quality.

Sampling

As for barley, malt, and adjuncts, one must analyze representative portions of the hops or hop-derived material being purchased. For unprocessed hops equal portions should be taken from five to ten places in a heap—different heights and depths—until about 200 grams are obtained. In all cases samples are sealed in a container and refrigerated because hops are tremendously susceptible to deterioration.

If the hops have already been compressed into bales, about 10 percent of a shipment under 100 bales or the square root of the number of bales if over 100 are sampled. 100 gram samples are cut from either end of the bales or alternatively (and preferably) such samples are taken from the heart of the bales using a device called an Oregon Sampler: a steel tube about 25 cm long and 7.5 cm in diameter with one pointed end and gripping handles at the other end.

Many brewers these days have their hops pelletized or converted into liquid extracts. Sampling is of 200 grams from every tenth box of pellets and a similar frequency for liquids.

All hops or hop product samples should be allowed to return to room temperature before examination and analysis. Hand inspection of cone hops is made on the sample directly, but for chemical analysis of all solid hop materials, they must be ground to a powder, taking precautions to avoid heat build up and attendant moisture loss. If the powder is not to be analyzed immediately, it is frozen.

Physical Examination Of Hops

Screening of cone hops is one of the most pleasurable things that can be done by rubbing. Brewers examine hop samples laid out on white paper as the basis for selecting hops from each year's crop.

A visual estimate of the degree of contamination of the sample with debris such as leaves and stems is made, perhaps reinforcing this quantitatively by picking the offending materials out from a 20 gram portion of hops and weighing them.

The color of the hops is a major clue to their condition, and the overall perception as dark green, olive green, pale green, yellowish green, or greenish yellow is noted. The relative proportion of seriously discolored cones in the sample (i.e., brown, red) is estimated. The luster of the sample is noted — are the hops glossy or dull? — as is the extent of cone breakage, if any.

The size of the cones is estimated, according to the general classification of large cones being 2¼ to 3 inches, medium cones being 1¼ to 2¼ inches and smaller ones being ¾ to 1¼ inches.

Ten cones are sliced in half longitudinally with a scalpel or razor blade, and the exposed lupulin glands are observed under daylight for an assessment of quantity, color (lemon yellow, orange yellow, or brownish) and degree of stickiness.

Then comes the rubbing. Several cones are rubbed between the palms of the (clean) hands and the whole raised to the nostrils. One is looking for desirable aromas (flowery is particularly appealing) but an absence of musty or cheesy character, which would indicate age or damage. (Incidentally the desirable notes that one is looking for don't translate simply to a given character in the ensuing beer. The purpose of this test is to check for the absence of nasty flavors and for the presence of characteristics that are representative of the desired varieties. Those characteristics are ones that the brewer knows will provide the aroma required in the finished product, albeit an aroma at variance from that which will be found in the hop itself.) A visual assessment of the level of seeds can be made here or from the 20 gram sample just referred to, which is dried, rubbed over a 20-mesh sieve, and weighed.

Aphid contamination of hops can be assessed using a relatively detailed boiling procedure involving borax and gasoline—not to be recommended for the non-specialist laboratory!

Chemical Analysis
The moisture of hops can't be accurately estimated by an oven-drying procedure, because the heat would drive off the essential oils as well and their weight would be estimated as water. Only if the temperature is relatively low (i.e., 140 °F (60 °C)), and a vacuum is applied should a drying technique be trusted. Otherwise distillation-based procedures are used.

The principal criterion (other than the findings of rubbing and visual inspection) by which hops are traded is their content of α-acids, the precursors of beer bitterness. The β-acids contribute to a much lower extent, after oxidation, so some people feel it useful to have an estimate of them as well.

The standard procedure in the United States is to extract freshly ground hops with toluene and to measure the absorbance of ultra-violet light by the resultant extract after dilution. Three wavelengths are used: 275, 325, and 355 nm. The content of resins is calculated using these fearsome-looking equations:

$$\alpha\text{-acid } (\%) = d \times (-51.56A_{355} + 73.79A_{325} - 19.07A_{275})$$
$$\beta\text{-acid } (\%) = d \times (55.57A_{355} - 47.59 A_{325} + 5.10 A_{275})$$

where d is the dilution factor.

The alternative approach is to titrate the toluene extract with a standardized solution of lead acetate and monitor the electrical resistance. As increasing quantities of lead acetate are added, the resistance of the mixture declines up till the point at which all the positively-charged lead ions are bound up with negatively-charged resins, after which resistance starts to rise again. The point of minimum resistance relates to the level of resin.

These techniques for assessing resins can be used with pelleted hops, though not pellets in which isomerization of resins has been deliberately promoted by the processor using warm storage. Evaluation of these isomerized hop pellets is accomplished by extraction with acidified butyl acetate and quantitation of the iso-α-acids by high performance liquid chromatography. HPLC is employed to assess iso-α-acids in pre-isomerized hop extracts. It can also be used to assess un-isomerized α-acids and β-acids in kettle extracts and solid forms of hops (after careful laboratory extraction of the resins with ether). Of course the resins might also be determined in such extracts using the spectrophotometric and conductimetric methods just referred to. The advantage of the HPLC technique is that information can be obtained on the individual types of α-acid or iso-α-acid. These differ to a certain extent in their bitterness or potential bitterness and in their foaming characteristics. HPLC is also used to assess the levels of reduced iso-α-acids in those preparations that are so treated in order to be used in beers packaged in light-penetrative packaging such as clear or green glass.

I must stress again that hops are like professors of brewing science: sensitive creatures, prone to deterioration. The resins degrade when oxidized, to produce whiffs of overly done Parmesan, elderly Stilton, or well-worn socks. The extent of deterioration can be assessed as the Hop Storage Index (HSI), which is computed on the basis of changes in the u.v. absorbance characteristics in the method I just referred to. During aging there is an increase in absorbance at 275 nm relative to that at 325 nm.

$$HSI = \frac{A_{275}}{A_{325}}$$

The assessment of the essential oil fraction of hops is even more challenging than that of the resins. The total oil component can be estimated by steam distillation, weighing the fraction of material from ground hops collected in the receiver. As yet there is no fully accepted method for estimating the individual components of the oil and relating those to desirable or undesirable attributes.

Exercise

1. Analysis of four hops yielded the following data:

	A	B	C	D
Seeds (%)	< 0.1	1.1	0.2	0.2
Moisture (%)	15.1	8.3	9.2	9.3
A_{355}	0.565	0.732	0.615	0.613
A_{325}	0.510	0.610	0.596	0.543
A_{275}	0.120	0.140	0.132	0.157
(dilution factor, d)	0.667	0.667	0.667	0.667
Total essential oils(%)	2.1	1.2	1.5	1.5

a. Which of these hop samples has received insufficient drying?
b. Which of these hop samples might you predict to be the aroma variety amongst them?
c. Which of the samples displays the greatest degree of deterioration?
d. Which of these might you predict to have originated in the UK?
e. Which sample has the highest bitterness potential?
f. Which sample contains the most β-acids?

Chapter Nine

The Brew House

In chapter 6, I was at pains to point out the importance of achieving homogeneity in the malt to be used in brewing. And throughout I have been stressing the significance of representative sampling. For we are pursuing command: having materials that we know we can trust and which will behave in predictable and controllable ways. In this chapter, we review brewhouse operations and issues that bear on the production of consistent product.

Milling

One of the strongest illustrations of this is the milling stage in the brew house. Fundamentally the more extensive the milling the greater the potential to extract materials from the grain. However, in most systems for separating wort from spent grains after mashing, the husk is important as a filter medium. The more intact the husk, the better the filtration. Therefore milling must be a compromise between thoroughly grinding the endosperm while leaving the husk as intact as possible. Except for mash filters, that is, where the husk is irrelevant to filtration so milling can be very fine. (See Table 9.1.)

Certainly if we are dealing with a lauter tun or mash tun based system, the degree of modification of the malt is critical. If the malt is of uniform modification (i.e., all the kernels have attained essentially the same degree of modification), it is possible to rig up the mill to generate a constant and optimized

Table 9.1————————————————————————————————————
Balance Of Particle Sizes Needed For Optimal Runoff
In Lauter Tun Or Mash Filter Based System

Sieve Mesh	What is Retained	% Needed for Lauter Operation	% Needed for Mash Filter Operation
1.27	husk	18	11
1.01	coarse grits	8	4
0.547	fine grits 1	35	16
0.253	fine grits 2	21	43
0.152	fine grits 3	7	10
residue	powder	11	16

particle size distribution. If the malt is uniformly well-modified, then the mill gaps can be relatively wide. They would be narrowed if the malt were uniformly less well-modified. However, if the malt is of uneven modification, it is not straightforward to produce idealized milled grist. Take for example a malt that is 80 percent very well-modified and 20 percent severely under-modified, perhaps because it was produced (carelessly!) from a somewhat dormant or dead barley. If the mill was set up to deal with the well-modified stuff then the under-modified kernels would receive limited break up, the particles produced would be very big and not readily extracted in the mash (apart from the large quantities of viscous β-glucan that they would release). If the mill was set up to deal with the coarser particles with very tight gaps this would pulverize the well-modified portion, generate large amounts of fine powder, and shatter the husk. This would also retard and even clog separation systems. Using more complex mills (such as 6-row mills) helps equalize such variations—but only to an extent. For mash filters, the grist is hammer milled to get very fine particles. Although this does equalize the particle size, any grossly under-modified corns will still release their β-glucan and wreak havoc downstream.

Mill settings are modified (regularly but not continually) in response to the quality of the grist. A finer grind is needed for malt of lower modification. The brewer will inspect the milled grist using standard sieves.

Mashing: Calculating the Grist

One of the purposes of the extract value for a malt or adjunct sample is to enable the brewer to calculate how much of that material is needed for the production of a certain volume of wort of the desired strength (gravity). Momentarily I will give some examples of how it's done. Let me mention first, though, that brewers in different parts of the world use different units here, just as they do for other parameters. We have already encountered the measurement of temperature as an example of this. In my opinion, brewers internationally should stick to a common currency. It would make life so much easier. Despite my Anglo-Saxon heritage, I won't speak of liter-degrees per kilogram as a unit of extract, unlike many of my compatriots. Equally, please don't expect me to pander to parlance such as brewers pounds per barrel or bushels of malt. If you really must seek your own favored units then you need to go to Appendix 1 for the relevant conversion factors.

Example 1

- You require 500 hl of wort at 15 °Plato.
- This needs to be produced from an all-malt grist, and the malt has an extract of 79%.
- To calculate the total amount of extract needed, multiply the volume by the °Plato value:

  i.e., 500 x 15 = 7500 (or 7.5 x 10³) hl degrees.

- Turning to the malt, had it been capable of total dissolution (100% extract, which is impossible but I am making a point!), we would have needed 7.5×10^3 kilograms (recalling that the definition of °Plato is on a percentage by weight basis).

- However, the malt is 79% extract. Therefore rather more of it is required to generate the extract we need—of course, we divide by 0.79.
- So the weight of malt needed is $7.5/0.79 \times 10^3 = 9.5 \times 10^3$ kg.

Example 2
- You require 1200 hl of wort of 19.5 °Plato.
- The recipe specifies that this should be produced from a grist of 60% malt, 32% corn and 8% sucrose. The malt has an extract of 76%, the corn of 81% and the sucrose 98%.
- The total extract required $= 1.2 \times 10^3 \times 19.5 = 2.34 \times 10^4$ hl deg.
- The extract to come from the malt is 60% of this, i.e., $60/100 \times 2.34 \times 10^4 = 1.4 \times 10^4$ hl deg.
- The extract to come from the corn is $32/100 \times 2.34 \times 10^4 = 0.75 \times 10^4$ hl deg.
- The extract to come from the sugar is $8/100 \times 2.34 \times 10^4 = 0.18 \times 10^4$ hl deg.
- The malt is 76% extract so we require $1.4/0.76 \times 10^4 = 1.8 \times 10^4$ kg.
- The corn is 81% extract so we require $0.75/0.81 \times 10^4 = 0.93 \times 10^4$ kg.
- The sucrose is 98% extract so we require $0.18/0.98 \times 10^4 = 0.18 \times 10^4$ kg.

When it comes to assessing how efficient the extraction has been during mashing and wort collection, the brewer calculates back according to the equation:

$$\% \text{ yield} = 100 \times \frac{\text{extract obtained in wort collected in the kettle}}{\text{extract available in the raw materials (x)}}$$

where x = the summation of the weight of each raw material multiplied by its percent extract yield as predicted in the laboratory mashes. In a perfect world the extract obtained should match that predicted from the earlier equations. Of course, utopia does not exist. While the brewer should not expect to be far off in achieving their anticipated yield, it is unlikely to be a perfect agreement. We must remember that small-scale mashes are hardly perfect matches for what occurs in the brew house. They are well-mixed systems, where the extract recovered after filtration is uniform throughout wort collection. By contrast, the extract obtained in a commercial brew house has much more variation through the collection into the kettle. First worts are naturally stronger, becoming successively "thinner" as sparging progresses. Furthermore, the ability to recover the extract is adversely impacted by any factor that retards wort separation, such as high levels of β-glucan. Naturally, one should expect a rather good agreement between extract values measured on incoming sugars and extract recovery from such materials in the production wort.

Mashing: Hitting the Correct Temperature

The grist is relatively cool: essentially it will be at ambient temperature. If you add hot water to it, the particles will tend to take heat from the water, and the net temperature of the mash will be reduced. Therefore if the correct mashing temperature is to be reached, the water must be at a higher net temperature than target and must be mixed with the milled grist during mashing so that the desired temperature is uniformly achieved for the entire mash. There is another complication, something called "slaking heat," which is the heat evolved when the extract is dissolved. Without getting hung up on this, let's just hit the necessary equation (which works in Celsius or Fahrenheit, but just don't mix the scales up!):

$$T_t = \frac{(S \times T_g) + (R \times T_w)}{R + S} + X$$

Where T_t = target starting mash temperature

S = specific heat of the malt grist (see Table 9.2)

T_g = temperature of the grist

R = ratio of water: grist (hectoliters/100 kg)

T_w = temperature of the water

X = slaking heat correction (see Table 9.2)

Table 9.2
Values Required For Computation Of Striking Temperature

Moisture content of the malt (%)	Specific heat of the malt grist	Slaking heat correction (°C)	Slaking heat correction (°F)
0	0.38	3.1	5.5
1	0.38	2.6	4.7
2	0.39	2.3	4.1
3	0.40	2.1	3.7
4	0.40	1.7	3.1
6	0.41	1.3	2.3
8	0.42	1.1	2.0

Example

We aim to mash in, at 149 °F (65 °C) and a water-to-grist ratio of 2.6, a malt which contains 2.5% moisture and is at 68 °F (20 °C). Therefore (and using Celsius) the target temperature of the striking water (T_w) is found from:

$$65 = \frac{(0.395 \times 20) + (2.6 \times T_w)}{2.6 + 0.395} + 2.2$$

or

$$65 = \frac{7.9 + 2.6\,T_w}{2.995} + 2.2$$

Rearranging

$$T_w = \frac{[(65 - 2.2) \times 2.995] - 7.9}{2.6}$$

Therefore the temperature of the striking water needs to be 69.3 °C (156.7 °F).

A similar type of calculation is also used for computing the temperature changes in decoction mashing systems and for predicting the conversion temperature when an adjunct mash is added to the malt mash in a double mashing system.

Controlling and Monitoring Mashing and Sweet Wort Collection

The foregoing shows that the principal control parameters for mashing are measures of mass, volume, and temperature. For many brewers the only test that will be applied to check that mashing has been successfully completed is to take a sample of the mash, put it into a well on a white plate, and add iodine. The absence of blue coloration indicates that starch has been dealt with.

Wort separation systems will be regulated by control of pressure differentials and raking of the grains (in lauter tuns) and monitored by checking (a) the specific gravity of the wort (in most bigger operations this is done with an in-line monitor); and (b) the wort clarity, in the latter case using an in-line haze meter operating on the basis of light scatter. Excessively turbid worts may be recycled.

Good extraction of soluble material from the grains demands sparging with water. The use of insufficient water will mean that extract is left behind in the grains: the cows that will feed on those grains will be fatter, but the brewer's wallet won't be. Equally, too much sparge water will lead to an excessively dilute wort, the risk of insufficient space in the kettle, the need for excessive boiling to try to drive off the extra liquid and, if the reduced gravity is carried forward to the fermenter, an atypical fermentation.

The rate of wort separation is described by Darcy's Law, which is described by this equation:

$$\text{rate of liquid flow} = \frac{\text{pressure x bed permeability x filtration area}}{\text{bed depth x wort viscosity}}$$

Interpreting the equation, we conclude that if the pressure applied to the top of the system is increased, this will increase the rate of liquid flow, rather like a plunger on top of a syringe. However beware, excess pressure will force the particles in a bed together, and this will tend to block the system.

One of the main determinants of bed permeability is particle size. The bigger the particles, the greater the permeability. The best analogy is to sand and clay (See Figure 9.1.): sand particles are bigger, and water can percolate between them readily. Clay particles are much smaller, and water tends to be retained by having to meander its way between them, as anyone who has tried to dig clay soil is aware. Hence the advantage of large husk particles when they are acting as a filter bed in lauter tuns. And this is the reason why the sticky layer that collects on a grain bed (the teig) with its compacted collection of insoluble materials is detrimental to wort collection.

The bigger the filtration area, the greater the rate of liquid collection—which is why lauter tuns are rather wide, and why mash filters have lots of plates.

An increase in any factor on the bottom (denominator) of the right hand side of this equation leads to a reduced rate of liquid collection. Obviously the longer the distance which the liquid has to travel (bed depth), the slower the wort separation will occur. Ergo lauter tuns tend to be shallow. Hence the

Figure 9.1
The Porosity of Clay and Sand

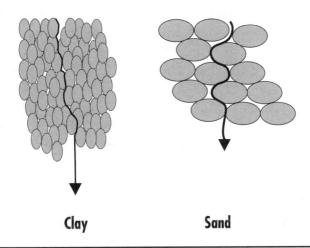

Clay   **Sand**

advantage of mash filters with many short beds in the form of the distance between separate plates.

Highly viscous liquids flow more slowly—think of the oil you pour into your engine. No wort I have seen approaches that degree of viscosity, but the presence of high levels of residual β-glucans will be sufficient to make wort separation sluggish.

Wort Boiling
Wort boiling is still regulated by many on the basis of percentage evaporation, i.e., the volume reduction during boiling. Alternatively the boil may be regulated solely according to length of boil. Perhaps the most significant calculation at this stage is that of the hop grist.

Taking as an example a beer hopped traditionally with cone hops, the important information we need is the α-acid content of the hops and information about the anticipated hop utilization for that brewery under the conditions to be employed.

$$\text{Hop utilization } (\%) = \frac{\text{iso-a-acids in beer x 100}}{\alpha\text{-acids added to the kettle}}$$

For cone hops the value may be relatively low, as poor as 25 to 30 percent. Impacting factors include pH (the lower the pH, the worse the utilization), gravity (lower gravities give better utilization), and anything that leads to loss of α-acids or iso-α-acids by adsorption, i.e., on trub or yeast.

Thus, if the utilization is expected to be 30 percent, the brewer knows that he needs to add 100/30 = 3.3-times more α-acid than would stoichiometrically yield the desired bitterness level.

Let's say the brewer is making 500 hl of a wort of "sales gravity", i.e., it will not be diluted after fermentation. The target bitterness is 25 Bitterness Units (i.e. 25 mg/l).

Plugging the values into our equation:

$$30 = \frac{25 \times 100}{\alpha\text{-acid}}$$

we rearrange this to:

$$\alpha\text{-acid} = \frac{25 \times 100}{30}$$

ergo, the amount of α-acid that needs to go into the wort is 83.3 mg/l. There is 500 hl of wort, so that is a total of 500 x 100 x 83.3 = 4, 165,000 mg α-acid, or 4.165 kg of α-acid.

Now our brewer has two hops at his disposal, one of 11.4 percent alpha and one 3.1 percent alpha.

Thus to get 4.165 kg of α-acid from the first hop, which contains 11.4 g/100 g (0.114 kg/kg), he would need to add 4.165/0.114 = 36.5 kg hops. To get the same yield of bitterness from the second hop, the addition would be 134 kg hops.

Note that in the latter case she would be adding more hop material in dry weight terms, material that is not resin. This will include oils, with attendant possible flavor effects, and more polyphenol, which will present an increased colloidal instability challenge.

One other factor frequently overlooked is that the boiling point of an aqueous system such as wort changes with altitude. The higher up you go, the lower the temperature at which water boils, according to the equation:

$$\text{Boiling Point (°C)} = [-1.07 \, (A \times 10^{-3}) + 100.1]$$

where A is the altitude (feet x 10^{-3})

Analysis of Wort

Considering wort is the fermentation feedstock for yeast, it is remarkable how little effort most brewers devote to its analysis. Whereas they worry long and hard about the yeast (See chapter 10.), just about the only measurement made routinely on wort is specific gravity. They assume that if this is at the desired level, all the components of the wort will be in the appropriate proportions. This is a naïve assumption. However, it is founded on their confidence in malt analysis: they believe that if a malt as evaluated in small-scale mashes delivers the appropriate mix of specific gravity, fermentability, free amino nitrogen, and pH (See chapter 6.), so it will when mashed on the commercial scale. Thus, just measure specific gravity, and all else will look after itself. This places too much trust on small-scale mashes, which are unrealistic in terms of milling, volume-to-surface area ratio, wort separation approach, and more. To repeat a point I made earlier, they tend to be well mixed, and the wort flowing through the filter paper at the completion basically reflects the average composition of all the wort in that mash. By contrast, on a commercial scale the properties of the wort change throughout run-off, being stronger at first, then getting progressively weaker. The composition changes in terms of relative balance of materials, buffering power, etc. Net wort composition will depend on the precise manner by which those worts are collected and distributed into kettles.

Once again, then, we must be aware of the significance of sampling if we are to make real sense of wort analysis from commercial breweries. Samples should be taken such that they properly represent the wort at the point of interest. They should be analyzed immediately, or else after cold storage. Worts show a tendency to become oxidized and to throw hazes, the latter depending on precise regimes of cooling. For meaningful comparisons to be made, sampling and sample treatment must be completely standardized.

The standard methods for assaying wort were mentioned in chapter 6.

Exercises

1. Here is analytical data on three worts:

	E	F	G
Specific gravity	1.0541	1.0732	1.0410
Real extract of fermented wort (°P)	4.54	6.26	0.51
pH	5.6	5.1	5.4
Free Amino Nitrogen (mg/L)	202	271	120
Viscosity	1.5	1.7	1.9
Zn^{2+}	0.03	0.3	0.2

a. Which wort was produced by very high gravity techniques?
b. How many °Plato in wort E?
c. One of these worts was produced with the inclusion of glucoamylase in the mash. Which?
d. Which of these worts is most likely to have been mashed-in without a low temperature stage?

e. Which wort might ferment sluggishly?

f. Which wort was acidified?

2. Assuming 1 Bitterness Unit equates to 1 mg/L iso-α-acid and you have 15 liters of wort, how many grams of hops (4.3% alpha) would you need to use to achieve a target bitterness of 15 BU. (Efficiency of utilization = 30%)?

3. You are mashing in 200 hectoliters of an all-malt grist at a water-to-grist ratio of 3 to 1. The milled malt (3% moisture) is at 68 °F (20 °C). The recipe demands a mashing temperature of 152.6 °F (67 °C). What is the temperature of the water you need to add?

4. You need to produce 1500 hectoliters of wort of 10 °Plato. The recipe stipulates that 75 percent of the final extract must be derived from malt and the rest from High Maltose Syrup (HMS). The malt at your disposal has a laboratory extract yield of 73 percent whereas the HMS yields 94 percent.

What proportions of the two grist materials will you need to use?

5. You have 500 hectoliters of sweet wort, and the procedures manual dictates that you achieve an evaporation rate of 7 percent. How much water needs to be evaporated off?

6. You have used 100 kg of hops of 4.0 percent alpha to achieve a bitterness level of 20 BU in 550 hectoliters of wort. What is the utilization rate?

Chapter Ten

Yeast and Fermentation

They don't take it for walks or buy it flowers but most brewers would consider yeast their best friend. For the most part they will not over-stretch it, so they replace it with new yeast after as few as five or six successive fermentations and seldom more than 10-15. (I can think of several companies, though, where yeast has traveled from generation to generation without "fresh blood" throughout the history of that company).

And so most brewers protect their strains. They're reluctant to share them with others, eager to keep them at the peak of condition, and unwilling to stress them out or overburden them.

I well recall the day when, as a young quality assurance manager, my microbiologist came in to say that he'd looked at the latest culture of lager yeast that had come to us from H.Q. and had concluded that it was actually an ale yeast. A call to central command set them scurrying to check and, sure enough, labels had been mixed. (Curiously, another of our breweries had already used the yeast to ferment some lager, seemingly successfully, but let's not get sidetracked!)

This just shows that methods are needed to confirm and type yeasts. In these days of forensics, this is readily achieved for yeast by using DNA fingerprinting protocols. However, there is a battery of techniques of more widespread accessibility that may be used to confirm whether or not the yeast in question is the one you want it to be. Such tests are designed primarily to give information on whether the yeast is a lager or ale strain but can go beyond that to indicate which strain one has within those broad categories.

I will be mentioning solid and liquid culture techniques. In the case of the latter I mean yeast growing in suspension in media that provide the necessary nutrients for growth, which may be monitored visually as turbidity or by counting or weighing the cells. Brewery fermentations are one large-scale example of this but in the laboratory such cultures may be performed on scales as small as test tubes. When talking of solid culture, I refer to yeast growing on the surface of media that have been solidified, typically with agar. The yeast is typically spread on the medium using a flamed metal loop, in such a way as to spread the cells thinly (the suspension of yeast will be diluted to allow this). As the cells grow and multiply, they form colonies that may be counted. Each colony represents one original cell.

Differentiating Ale and Lager Strains

Lager yeast strains will grow on and ferment the sugar melibiose (which comprises a meld of galactose and glucose) whereas ale strains won't. If a yeast will ferment a liquid medium containing melibiose as the sole source of carbon and energy (as detected by gas and acid production, which causes a color change in a pH indicator), we can be confident that the strain is of the lager variety. An alternative procedure for performing the test is to incubate yeast on a solid medium in plates that incorporate X-α-Gal, a substrate for the enzyme that splits up melibiose. If the enzyme is present it will chop up X-α-Gal, and a green colony ensues. All that remains for the analyst to do after a few days is to see if the colonies are green. If they are, the yeast is a lager strain. If the colonies are white, it is an ale strain. If there is some of each, there is a mix of yeasts, probably because of contamination.

A method that's simpler than these methods is to plate out yeast in two Petri dishes and put one into an incubator at 77 °F (25 °C) and the other into one at 98.6 °F (37 °C). Ale strains will grow (as colonies) at both temperatures, whereas lager strains grow only at the lower temperature.

Two of the more traditional techniques used to glean more information about the nature of a yeast strain are assessment of giant colony morphology and flocculation. When yeast is grown for three to four weeks on plates incorporating wort solidified in gelatin, they may produce giant colonies of weird and wonderful shapes that are strain characteristic. When suspensions of yeast are made in calcium sulfate solution at pH 4.5, they will settle out to varying extent depending on their tendency to aggregate and flocculate. This can be quantified on the basis of light scatter at 600 nm.

Pitching Yeast

Whether the yeast is derived from a propagation or from the proceeds of beer making in the brewery, it is essential that it fulfills several criteria: it should be the right strain, contamination free, healthy and in the correct quantity. I will concentrate on yeast collected after fermentation for re-use, but the same considerations apply to yeast at the conclusion of a propagation exercise.

Yeast slurries and suspensions become stratified, and it is essential to keep that in consideration when sampling. If sampling is from a storage tank, the contents should be roused. If samples are taken as yeast is being transferred by pumping, samples (100 ml) should be taken at intervals, followed by mixing of the samples prior to analysis. Yeast should be assessed as rapidly as possible and held at less than 35.6 °F (less than 2 °C) pending completion of the analyses. Handling techniques must be aseptic.

Just as for barley, malt, and hops, the eyes and nose are valuable analytical tools when it comes to yeast. Does the sample smell good or is it whiffy? Does it taste good? (I know of one company that tastes everything that comes anywhere near the product with the possible exception of licking the floor.) Are its color and tendencies in suspension (flocculent or not?) as expected for that yeast?

The visual inspection should always be further facilitated using a microscope. Do the cells look good and healthy? Are there any abnormal looking beasts? Any sign of alien bugs?

One cannot easily tell if cells are alive or dead by simple microscopic examination. The handed down method for this invokes the use of methylene blue—although some argue for the superior attributes of methyl violet. These are stains that are decolorized by living yeast. If a culture of yeast as viewed through the microscope is colorless after staining, it is alive. By counting the number of blue or violet cells as a percentage of the total cells present then an estimate of percentage viability is obtained. If the number of stained cells is too large (e.g., > 10-20%), the yeast should not be used.

An alternative approach is to plate yeast on tiny blocks of agar medium located on microscope slides. Percentage viability equates to the proportion of cells that give rise to micro-colonies in relation to the total number of cells put onto the slide.

Many brewers these days talk about vitality in addition to viability. I heard a somewhat vulgar explanation of this once, but permit me a more polite one. Both baseball player Randy Johnson ("The Big Unit") and I are viable, but can you tell me which of us is the more vital and healthy? In the same way, yeast may be alive but hardly kicking (or to extend the Johnson analogy using a cunning play on words—fit for pitching). It is only when yeast is in the peak of condition that tiptop fermentations will occur. Various methods have been proposed for assessing vitality of yeast but none is as yet accepted globally as providing sufficiently reliable numbers. The machines that measure capacitance are showing the most promise. It behooves the brewer therefore to apply a quality assurance mentality to the problem: look after the yeast, and it will naturally display sufficient vitality so there will be no need to make measurements on it. Don't leave the yeast sitting around on the beer at the end of fermentation because it will start to autolyze, leading to damage of yeast and beer alike. Keep the yeast cool between fermentations. Don't use it too many times.

Assuming we have healthy yeast, the next requirement is to pitch it in the correct quantity. This can be done on the basis of weight or number. They will not necessarily throw up the same answer because yeast cells can differ in size. Various factors will influence the size of a cell, including its state of health, age, and nutritional status. If yeast quantitation is done on the basis of number of cells, size is immaterial. If, however, measurement is on the basis of weight, if the average cell size increases, fewer cells would be pitched at a given weight.

Yeast counting is performed using a microscope slide featuring a grid of squares. The yeast is diluted by a known amount such that individual cells can be observed on the slide. The number of cells in several squares is counted and multiplied by the dilution factor to obtain the number of organisms in the original culture.

We can quantify yeast-solids in two ways: by dry weight determination and by a spin-down procedure. The former procedure is much like that performed for barley, malt, adjuncts and hops for assessment of moisture. The risk is that any non-yeast mass will confound the result so things like trub must first be removed by sieving through a 100-mesh sieve. The weight is assessed after several hours of drying.

The spin down procedure involves centrifuging some yeast slurry in a graduated tube. The slurry will divide into yeast and trub layers, and from this the percent (by volume) of yeast solids and total solids can be estimated. Methods such as this are often calibrated against either weight or cell number information so the volume reading can be related to a pitching value on the basis of weight or number. In more recent years, instrumental procedures based on light scatter and capacitance measurement have come on to the market, and these are calibrated against the more traditional procedures.

A typical pitching rate rule of thumb is 1 million cells per milliliter per degree Plato.

The other important ingredient that the yeast needs in order to efficiently perform its fermentation is oxygen. This confuses a lot of folk because alcoholic fermentation is an anaerobic process. The yeast, though, needs the oxygen in relatively small quantities to fuel the production of important components for its cell structure. The majority of brewing yeast strains are satisfied if the wort is either saturated with air or saturated with oxygen. Air of course is only 20 percent oxygen (the balance is for the most part nitrogen), so basically a yeast for which air saturation is sufficient needs about a fifth as much oxygen as does one that demands oxygen saturation.

Monitoring Fermentations

The standard approach to monitoring fermentations is to follow the drop in specific gravity that occurs as the high density sugars are converted into ethanol, which has a density considerably lower than water. Other parameters can be measured: such as the release of CO_2, or the drop in pH, or the production of yeast, or the formation of ethanol. For the most part, monitoring the gravity drop and adjusting conditions in response, notably temperature, can enable sufficient control. If the fermentation is running too quickly, cooling may be applied (assuming the fermenters are jacketed). If running slowly, the cooling may be switched off and the temperature allowed to rise as a natural consequence of yeast metabolism. However, most brewers would strive to ensure that the conditions at the start of the fermentation are on specification (viable yeast count, the appropriate oxygen charge, addition of zinc at about 0.2 ppm).

There is one other critical measurement—diacetyl (and perhaps another so-called vicinal diketone, pentanedione). These heinous substances cause beer to reek of butterscotch or honey. They are produced naturally as an offshoot of metabolism but the yeast will mop them up again. Yeast do this only if healthy and given time to do the job. The responsible brewer should measure vicinal diketones, or VDKs, as well as their precursors. Since the precursors break down when heated, the best method is to use gas chromatography to quantify the diacetyl and pentanedione and then repeat the analysis after the sample has been heated. The important value is total VDK, which is the precursors plus the diacetyl and pentanedione. If the free VDKs are low but the precursors remain, this is unsatisfactory because those precursors will sooner or later break down to give lousy flavor in the beer. For those with no gas chromatograph, VDK can be assessed by color generation with α-naphthol. However, even this is not a straightforward procedure. The smallest brewing operations might need to resort to classic QA: look after the yeast, get the fermentation conditions absolutely right, don't rush the beer through the system, and for goodness sake avoid contamination—some bacteria are prolific producers of diacetyl. (In fact, for those with the analytical rigs, the ratio of diacetyl to pentanedione can be used diagnostically. Increases in the ratio signals bug problems.)

Exercises

1. You have two yeast slurries, both of which contain the same weight of yeast, but the second of which contains 10 percent more cells. Explain the difference.

2. To achieve consistency in fermentations using the yeasts in question 1, would you pitch on the basis of weight or cell number?

3. Which of the following is an ale strain, which a lager strain, and which is a mixed culture?

	A	B	C	D
Growth on melibiose	+	−	+	+
Growth at 37 °C	−	+	+	−
Colonies on plates incorporating X-α-Gal	Green	White	Green, white	Green

4. Discuss the possible reasons for these different shaped lines obtained in three separate fermentations with the same wort run from the lauter tun and the same yeast.

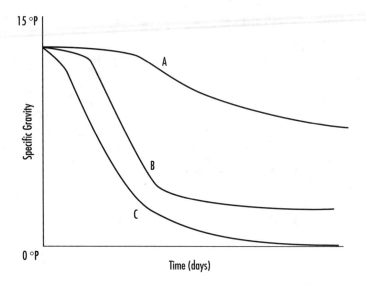

5. For the fermentations in question 4, which will produce the most alcohol?

6. You need to pitch 500 hectoliters of wort with viable yeast at a rate of 12 million cells per ml. In a methylene blue check 3 percent of the cells appear blue. You have a slurry of yeast that has a consistency of 5 percent solids. Using the rule of thumb that 10 million cells/ml approximates to 0.3 kg/hl, what volume of yeast slurry will you be pitching into your fermentation?

Chapter Eleven

Beer

By now even the reader who was totally unfamiliar with brewing quality control before picking up this book should have learned that many analyses are made on the raw materials and process stream in pursuit of the controlled production of a consistent beer. It's now time to analyze the finished product, recognizing that if the efforts in pursuit of control were well done, many of these checks should merely confirm the quality.

Many of the samples will be of beer in final package. Representative packages should be analyzed—i.e., take samples at various stages in a packaging run. Sometimes it is desirable to take samples from a can or bottle under aseptic conditions. Prior to this kind of sampling, flame the opening before removing beer with a sterile pipette or pouring the sample into a sterile container. For the most part this won't be necessary, but it will be when you're sampling from process lines and tanks in order to avoid contaminating the process stream. Sample ports must be scrubbed, swabbed with alcohol, and flamed before running enough beer to cool the metal down prior to collecting the desired quantity.

For many analyses, beer needs to be decarbonated by gently shaking it in a conical-shaped flask. Alternatively the beer may be agitated on a mechanical shaker. There are also filtration techniques available that allow rapid removal of gas. In all cases the beer should be presented for subsequent analysis at a temperature around 68 °F (20 °C).

Chemical Analyses

Perhaps the most focused upon analyses made on beer are those of alcohol and specific gravity. From these the brewer determines whether the beer is of the appropriate strength and has fermented properly. In some countries excise tax is levied in proportion to the alcohol content, making it important for cost reasons to measure it to a precision of 0.1 percent or better.

Several procedures can be used to measure ethanol. The alcohol can be collected by distillation, and the quantity measured by volume or weight. It can be assessed using refractometry or by gas chromatography. Instruments like SCABA measure ethanol by catalytic combustion. Specific enzymes can be used for measurement of ethanol for low alcohol products.

The specific gravity of beer can be assessed using several instruments, notably the pycnometer, the hydrometer, and the digital density meter. Operating on the principle of a vibrating U-tube, the digital density meter is widely used. The frequency of oscillation is tempered depending on the relative density of the liquid in the tube and is directly read off in an illuminated display on the meter.

The net specific gravity of beer is influenced by the presence of alcohol, which has a specific gravity of 0.79. The more alcoholic the beer, the lower the measured specific gravity at a given concentration of residual carbohydrate. Thus values for residual extract made on beer are generally referred to as apparent extract. If the measurements are made on the residue after removal of ethanol, the value is called real extract.

Using the measurements of real extract and alcohol, it is possible to calculate several pertinent parameters.

Original Extract

By knowing how much alcohol is present in beer and how much extract is left behind in the beer, we can calculate what the strength of the wort must have been at the start of fermentation, the original extract. The formula is:

$$O = 100 [(2.0665A + E)/(100 + 1.0665A)]$$

where O = original extract (°Plato), A = alcohol (% by weight) and E = real extract (%).

Real Degree of Fermentation (RDF)

The RDF is the extent to which fermentation has genuinely occurred. It is calculated by another scary looking formula. Don't worry about it, just accept it:

$$RDF (\%) = \{[100(O-E)]/O\} \times \{1/[1-(0.005161 \times E]\}$$

O and E are defined in preceding paragraphs.

Caloric Content

One last value determined by calculation is that of caloric content. The calories in beer are located in carbohydrate, protein, and alcohol. The generally accepted values are 4 kcalories (kcal) per gram for carbohydrate and protein and 6.9 kcal/g for alcohol. The relevant equation is:

$$\text{Calories (kcal/100 g beer)} = 6.9(A) + 4(B-C)$$

where A = alcohol (% by weight), B = Real Extract (% by weight) and C = Ash (% by weight).

We have already encountered some of the other methods used on beer when we discussed malt and wort analysis. Parameters such as protein content, free amino nitrogen, viscosity, nitrosamines, and pH are all measured similarly in beer.

Color is also measured in beer and wort in comparable ways. Because most beers are bright, however, there is no need for a filtration step prior to spectrophotometry. Whilst the simple measurement of light absorbance at 430 nm appears to be adequate as an index of color measurement for the paler lager-style beers, it seems to be generally inadequate for darker products. After all, color results from absorbance or lack of absorbance at more than a single wavelength. The more complex the grist (i.e., use of colored specialty malts) the greater the range of colored materials present. Dark beers diluted to have the same color as a lager (as judged on the basis of A_{430}) are perceptibly different—perhaps pinker.

One component present in beer but not found in wort is carbon dioxide. The simplest way to describe measurement of CO_2 is to say that it is done by using a CO_2 meter! Traditionally CO_2 is measured manometrically: the gas is trapped as sodium carbonate by reacting the beer with caustic and it is hooked up to a manometer (a device that measured gas pressure, like a barometer). The CO_2 is then released by adding acid: the higher the pressure registered, the more CO_2.

A gas never desired in beer is oxygen as it leads to staling. Instruments based on the Clark electrode are available for off-line and on-line assay of O_2, which can also be measured using indigo carmine, which develops a blue color when reacting with O_2.

The bitterness in beer is most readily assessed by extracting the bitter acids (iso-α-acids) in iso-octane and measuring the absorbance of the resultant solution at 275 nm. The relevant equation is:

$$\text{Bitterness units (BU)} = A_{275} \times 50$$

More information about the individual iso-α-acids is obtained by use of HPLC to fractionate the beer. This is certainly the way to measure bitterness if the bittering is by specialized hop preparations such as the reduced iso-α-acids.

Of the other major flavor contributors, the volatile materials (i.e., esters, sulfur compounds, etc.) are best assayed using specific gas chromatography based procedures. For each class of compounds there will be a different column, detector, and set of chromatography conditions.

One school of thought holds that certain inorganic ions are very important for beer flavor. Notably brewers talk of the chloride-sulfate ratio. There is a paucity of scientific justification for holding such a ratio in critical esteem but if you are convinced of its importance, the best analytical tool is liquid chromatography. With regard to the positive ions, those that worry me most are iron and copper because they vigorously promote staling and, in the case of iron, lead to metallic flavors. These and other cations are best assessed by atomic absorption spectroscopy.

Apart from flavor and color, two other key quality criteria are clarity and foam. The clarity of beer is easily checked by standing the beer in a clean glass against a well-lit background and looking at it. The edges of a black solid line located behind the glass should appear sharp and crisp when viewed through the glass.

A quantitative value for beer haze is obtained by measuring the amount of light scattered. A range of instruments is available—some measure scattered light at an angle of 13 degrees to incident and others at 90 degrees. (See Figure 11.1.) The physics is too complex to get hung up on, and it's best to just accept that the former type of instrument tends to exaggerate very large particles. The latter gives high values for very small particles that are difficult to see by eye (invisible haze). However, small particles grow into big ones with time, so the most sensible option is to make measurements with both types of instrument. The 90 degree meters in particular are used in breakdown studies. These tests accelerate the natural aging process in test samples so a view can be taken on whether a beer is likely to throw a haze over time. There are many variants of accelerated aging procedures, with a typical one having cycles of 24 hours holding the beer at 32 °F (0

Figure 11.1
Light Scatter

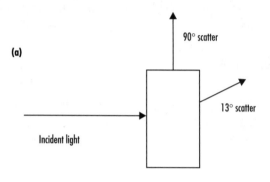

(a)

90° scatter

13° scatter

Incident light

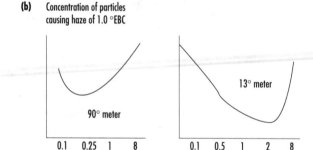

(b) Concentration of particles
causing haze of 1.0 °EBC

90° meter

13° meter

0.1   0.25   1   8   0.1   0.5   1   2   8

particle diameter (mm)

°C) followed by 24 hours at 98.6 °F (37 °C). One complete round at the two temperatures is said to equate to one month of natural storage in trade. Clearly this is a gross generalization.

Rather more useful would be to measure the levels of haze-forming components in the beer. The principle materials are protein and polyphenol, but only a portion of each is haze forming. The total protein level of beer is unlikely to simply relate to haze risk. Polyphenol measurements are more worthwhile—and can be assessed colorimetrically by reacting the beer with iron in alkaline solution and measuring the red color at 600 nm.

There are numerous methods for assessing foam in beer—confirming the fact that none are entirely reliable. The only one recommended by a brewing organization (ASBC) is the sigma value (Σ) method, which basically measures the rate at which liquid drains from foam. The more slowly this happens (higher value in seconds), the more stable is the foam.

Extreme foaming, otherwise known as gushing, can occur. One cause is if you live with a prat who shakes your beer cans when you're not looking. Solution: change roommate. The more frequent cause, however, is barley infected with Fusarium. (See chapter 5.) Solution: only use malt from uninfected grain. Test? The standard check for gushing in packaged beer is to subject the container to a standard shaking regime (one company swears by driving the beer over the city bumps in a truck) before opening the package (after a rest period) and measuring what weight of beer is lost.

Which leaves me with just one chemical parameter to mention: sulfur dioxide. It is a superb protectant against beer staling but it must be labeled for beers in the United States that contain more than 10 ppm. It is easily measured by reaction with p-rosaniline—a lovely blue color is produced that is assessed by measuring A_{550}.

Microbiological Methods
Beer is pretty much resistant to bug growth (There are some places I go where I ought to scrub my teeth with it rather than use the local H_2O.). This reflects a number of properties of the beer such as presence of antiseptic hop bitter acids and ethanol, low pH, and relative absence of nutrients. Beer is not a totally alien environment for bugs, and some will thrive in it. They need to be detected before they take a stranglehold.

Wort is rather more susceptible to spoilage than beer and needs to be checked too. Let's not forget the habitats where bacteria and fungi might be lurking in the brewery: water supplies, air, on the surfaces of tanks, and in pipes. Establish methods to check each of these.

As ever, we need to ensure representative sampling of what is present and to conduct sensitive testing to detect what may only be small numbers of organisms. Water, wort, and beer samples will be collected by the sterile procedures referred to earlier. (It is important that the procedure is sterile so we won't be measuring bugs introduced in the sampling protocol.) Trapping the organisms and growing them in the appropriate media will test air samples.

Traditional microbiological practice is to spread samples onto solid media in Petri dishes, incubate them either aerobically or anaerobically (in the latter case to detect bacteria which thrive in the absence of oxygen, remembering that beer is essentially anaerobic) and count the numbers of colonies developing after three to seven days. Samples containing relatively few organisms will be concentrated by first trapping the organisms on filters. A range of media is used to support the growth of the organisms in these dishes. Some of the media are general and used to determine the overall hygiene status of the sample. Others are more specific and used to detect specific classes of organisms.

Perhaps the most famous growth substrate is Wallerstein Laboratories Nutrient (WLN) medium, developed in that great brewing consultant's labs on Madison Avenue in New York City. It comprises glucose, yeast extract, hydrolyzed casein (a protein from milk) and salts, solidified in agar. It will grow pretty much all the organisms that will spoil beer. By adding nystatin, you prevent the growth of yeasts. Such a medium will be used to detect bacteria. Bugs growing on plates that don't contain nystatin may be bacteria or yeasts. Another way to do this is to grow organisms on a plate containing the amino acid lysine as the nitrogen source. Most spoilage yeast will use it and grow, but bacteria won't.

Raka Ray is not a flat fancy fish, but a medium used for detecting lactic acid bacteria (those bugs that spew out lactic acid and make the beer sour). Barney Miller medium, named for Mike Barney from the Miller Brewing Co., is another medium for detecting such organisms.

Another important test for figuring out what organisms you have is the Gram stain, named after a bloke called Christian Gram, the reason why you should always make it a capital G. In this procedure some organisms stain blue with crystal violet, the so-called Gram positive organisms. Those that don't stain blue are the Gram negatives. Bacteria such as lactic acid bacteria are Gram positive. Wort spoilers like the *Enterobacter*, beer spoilers like *Zymomonas*, and those that produce acetic acid (Yep, they're called acetic acid bacteria.) are all Gram negative.

More rapid procedures give data on hygiene status within a few hours. The most frequently used of these measure ATP and take advantage of an enzyme extracted from the rear ends of fireflies. This enzyme converts a substance called ATP into light that is emitted from the hind quarters to attract the opposite sex. (It never worked for me—the wife has always preferred a bunch of flowers.) In the ATP-bioluminescence test, swabs taken from tanks or samples of beer (usually after concentration on membranes— see earlier) are mixed with the enzyme. The ATP present in any organisms that are there is converted into light that can be detected in a photometer. The more bugs, the more light.

Quality Assurance in the Packaging of Beer

The filling of beer into containers—whether casks, kegs, bottles, or cans—can be a highly intensive stage in QC terms. It needn't be, provided proper QA mentality holds sway in decently designed and operated facilities. Once again the principle of piling as much responsibility as possible on suppliers is an important one. They should have to conform to and qualify in pre-agreed standards that will be periodically audited. Larger brewing laboratories will run a range of checks on the incoming raw materials (i.e., can color, bottle integrity, label dimensions, etc.) at a sensible frequency. On the lines the fillers will be checked head by head for parameters such as can-lid seams, torque on crown caps on bottles, and whether containers have received the right level of contents. And, of course, the integrity of the microbiological sterilization processes needs to be assured. Good brewery practice should focus on ensuring that the process is as clean as possible from start to finish so as few bugs as possible need to be ironed out. But some form of final microbial elimination strategy is generally necessary. Tunnel pasteurizers are checked using recorders that record temperature at various parts in the chamber. Those brewers using filters to eliminate microbes screen the integrity of filters, including challenge tests with microorganisms.

Sensory Methods

At the end of the day it's the smell and taste of a beer that determines its acceptability. In properly run breweries, people will do more than just taste the beer fresh into package. They will also taste raw materials and process streams, so any flavor defect is detected before the defective material has passed to the next stage in the process. This additional attention can avert wasted effort and dollars. In other words, it is classic QA. At the very least, beer should be tasted before and after filtration and immediately after packaging. I spoke earlier about tasting water, worts, etc. Tasters should be sensitive to all flavors that should and should not be present in each product, so several tasters should make decisions on samples. The head brewer may be almighty but it is by no means definite that he rose to his exalted rank because he was a better taster than anybody else. People differ considerably in their sensitivity to different flavors. People can be blind to some characteristics or acutely sensitive to others.

Beer tasting can be much more sophisticated than just having the head brewer and QA manager standing around a spittoon. Reliable and statistically well-founded tests that can provide meaningful information to enable decisions about beer quality are available. These procedures are divided into difference tests and descriptive tests.

Difference Tests

Difference tests tell if a difference is detectable between two beers. We might want to know whether a beer from one brewery is significantly different from the same beer brewed in a different location. We might want to check whether a process change has affected the flavor of the product. And so on.

In taste testing, all factors other than the taste and smell of the beer must be eliminated as distracting factors. The tasting room must be quiet, and the appearance of the product should be disguised so differences in foam, color, or clarity do not distort the picture. The beer is served in dark glasses and is one of the more respectable activities performed in a room bathed in red light. Cigarettes, coffee, burritos or other strongly flavored stuff should not have sullied the palate of the tasters.

The classic difference procedure is the three-glass test. Assessors are presented with three glasses, two of which contain one beer and one that contains the other beer. The order of presentation is randomized. The taster is asked to indicate which beer she thinks is different. Statistical analysis reveals whether a significant number of tasters can detect a difference between the beers.

Descriptive Tests

No training is needed to perform the three-glass test. However, if specific descriptive information is required about a beer, trained tasters are needed. These people are painstakingly taught to recognize a plethora of flavors and to discuss beers objectively and authoritatively to be able to profile a beer. (See Table 11.1)

Table 11.1 ━━━━━━━━━━━━━━━━━━━━━━━━━━━━━━━━━━━━━━━

Individual Flavor Profile Record Sheet

	Beer A			Beer B			Beer C			Beer D			Beer E		
	Ar.	Ta.	Af.	Ar.	Ta.	Af.	Ar.	Ta.	Af.	Ar.	Ta.	Af.	Ar.	Ta.	Af.
Alcohol															
Astringent																
Bitter																
Body																
Burnt																
Carbonation																
Cardboard																
Cheesy																
Cooked																
Vegetable																
Diacetyl																
DMS																
Estery																
Fatty acid																
Floral																
Fruity																
Grainy																
Grassy																
Hoppy																
Lightstruck																
Malty																
Medicinal																
Metallic																
Musty																
Phenolic																
Rancid																
Ribes																
Soapy																

	Beer A	Beer B	Beer C	Beer D	Beer E
	Ar. Ta. Af.	Ar. Ta. Af.	Ar. Ta. Af.	Ar. Ta. Af.	Ar. Ta. Af.
Sour					
Spicy					
Sulfidic					
Sulfitic					
Sweet					
Toffee					
Worty					
Yeasty					

Ar. = aroma; Ta. = taste; Af. = After taste
1 = Slight   2 = Significant 3 = Marked 4 = Strong 5 = Excessive

Exercises

Three beers were analyzed as follows:

	H	I	J
Alcohol (% by weight)	3.6	5.5	3.7
Real Extract (% by weight)	4.6	1.2	5.3
pH	4.1	4.6	4.3
Color (°)	7.1	3.9	5.2
CO_2 (vol)	2.1	2.6	2.7
Iron (mg/L)	0.1	0.4	0.05
SO_2 (mg/L)	9	3	11
Σ (sec)	134	132	91
BU	28	8	15
Diacetyl (mg/L)	0.08	0.07	0.13
Haze on storage (FTU)	130	270	141
Oxygen (mg/L)	0.1	0.15	0.08
Total polyphenols (mg/L)	161	95	122
DMS (μg/L)	38	36	62
Ash (% by weight)	0.6	0.2	0.5

a. Two of these beers were produced from the same wort of 12 °P. Which are they?

b. How might one of them have been processed differently in the fermenter?

c. Which beer has had its acidity adjusted artificially?

d. Which beer is probably intended for packaging into kegs?

e. Which has probably received a dose of colorant?

f. What other addition might this beer have received?

g. Which is likely to be most susceptible to oxidative staling?

h. Why else might this beer have the highest haze value after storage?

i. Which beer has received inadequate conditioning?

j. Which beer might display the most lager character?

k. Which beer has the most stable foam?

l. What would need to be declared on the label of one of these beers?

m. How many calories (Kcal/100g) in each of these beers?

Chapter Twelve

Approaching Quality Assurance for Big Guys and Little Guys

Years ago, as a fledgling QA manager, I was introduced to the Wednesday afternoon meeting. It comprised the packaging manager, his aide-de-camp, me and my own assistant. The latter, with sharp pencil and acid tongue, would draw our attention to page seven, column four, line eighteen and, jabbing his finger at the thick wad, would pronounce, "There, you see, a week ago today, bottling line 2, you had a million counts on filling head number six." I was supposed to look grave, glare at my compatriot on the management team and say sternly, "Well, what have you to say about that, then?" He would turn to his sidekick who would shrug his shoulders and say, "Dunno, but I've checked, and the beer has already been drunk in Wigan." Needless to say I quickly threw out the confrontational approach, utterly convinced that the wise brewer will place far more effort in the proactivity of quality assurance rather than the reactivity of quality control. Wherever possible, the only checks that will be made routinely are the ones that are necessary for the process to unfold. These will be made by the people who are actually operating the process, to avoid confrontational situations between those on the shop floor and the truth police from the lab.

In the case of the above example, which really did happen, we are forced to admit that when the beer really is that fast in getting from filler to bladder, the demands on it pertaining to shelf life are quite possibly less stringent than for a beer that will have a more leisurely lifespan. Another example of specifications being fit for purpose.

The ultimate ethos, whether you're an international brewing conglomerate or brewing in a bucket, should, as we saw in chapter 2, be "right first time". You should have systems and procedures that *assure* quality. Significant expenditure on systems designed to guarantee consistency should lead to much greater savings through avoidance of process failure and product rejection. (See Figure 12.1)

Another golden rule worth repeating is that if you are not prepared to respond to a measurement, don't perform it. Unless a piece of data is required and useful there is no value in generating it. It costs money to perform analyses. If all you're going to do is destroy more trees by ledgerising stuff and ignoring it then don't do it. It's a waste in every sense.

If raw materials are properly selected and used in an appropriately controlled manner, there is a strong likelihood that the product will be within specification. We must be realistic, however: brewing is based on biological entities that are inherently variable and unpredictable. Yet, as we have seen, it is perfectly possible to minimize the impact of this variation and to identify those measurements that provide worthwhile information on the extent of the deviation and on what can be done about it.

Bigger brewers will make many analyses in-line (i.e., using a probe directly located in a pipe or vessel) as part of feedback control systems. The next best option is an on- or at-line system, with the person responsible for the process step making the analysis and responding to it directly. For those brewing in buckets, this means *you*!

One more truism: if at all possible, pass the analytical buck. To my mind there is no point in any brewing company trying to measure everything in sight. Suppliers (whether of

Figure 12.1

Cost Savings From Instituting
A Pro-Active Quality Assurance Regime

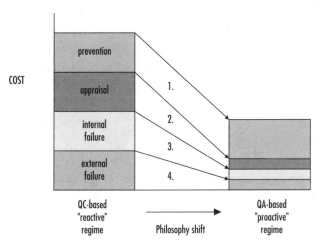

"Prevention" represents the cost of introducing systems to help assure quality, such as auditing, projects, in-line systems, training, awareness regimes, interactions with suppliers, engineering and installations etc.

"Appraisal" represents Quality Control monitoring costs – people, time, re-checks, external laboratories.

"Internal failure" equals cost of re-work, i.e. "Wrong first time" as opposed to "Right First Time" – e.g. reduced yields, re-filtration, blending, gas adjustment, scrap, trouble-shooting.

"External failure" amounts to complaints, lost markets, recalls, product liability, legal costs etc.

1 = cost of instituting pro-active QA regime
2 = savings on appraisal costs
3 = savings on internal failure costs
4 = savings on external failure costs
(2+3+4) – 1 = benefit of pro-active QA regime

malt, adjuncts, water, process aids, bottles, etc.) should provide the data, properly certificated, for perusal by the purchaser. The brewer should be able to have faith in the extract and moisture values supplied with the malt and be able to calculate the grist bill on this basis. Similarly the α-acid value in the hop must be reliable enough to allow hopping rates to be computed with confidence. The brewer should be auditing the supplier to assure satisfaction with the plant and processes. Equally it would be logical to perform spot checks of suppliers, to check agreement on reported values. Discrepancies occur for one of three reasons: a sampling inconsistency, an inadequacy in the analytical methodology as applied either by supplier or recipient, or a shortcoming in the material per se. These spot checks could just as easily, and probably more economically, be performed by a third party, a specialist lab whose product is analytical data rather than beer.

And certainly a degree of confidence would be introduced by supplier and third party analyst alike being high-performing participants in formalized ring analysis groups. Here, collections of laboratories perform analyses of samples distributed from a central resource to confirm that they are capable of getting data in good agreement with one another. In other words they are like the groups testing out new methods, only here they are using tried and trusted procedures that are expected to give good consistency. If they don't, in any given lab, they will have the finger pointed at them. Such rings exist within the bigger companies, with the labs in each of their breweries interacting in exercises coordinated from the central headquarters QA laboratory. These rings are also operated under the auspices of institutions. Based in the United Kingdom, but with an international clientele, there is the Brewing Analytes Proficiency Scheme (BAPS) administered by LGC (Teddington) Ltd. and Brewing Research International. The American Society of Brewing Chemists (ASBC) runs a check sample service. There are four series (beer, hop, malt, and barley) and, in each, samples are regularly distributed to participants who perform analyses, send the results to St. Paul (that's the HQ of the ASBC as opposed to the geezer who walked to Damascus) and then receive a confidential report of how they stand against all the other labs (c.f. my description in chapter 4). Actually the report tells them if their analytical methodology is passing muster. It's not especially cheap for the guy brewing in a bucket—the beer series in which one 6-pack is mailed for analysis four times a year costs $525 as I write. When balanced against the cost of potential quality failure that will surely result from not performing methods correctly, however, it is a small price to pay, especially if the participant is a lab that is set up for measuring all the parameters in the scheme (specific gravity, foam, air, color, bitterness, ash, iron, copper, sodium, calcium, carbon dioxide, sulfate, chloride, oxalate, total acidity, pH, real extract, residual fermentable extract, alcohol, reducing sugars, diacetyl, and protein).

The Basics

The list I have just given is quite a formidable one. And in the various chapters of this book I have listed all manner of tests that can be made right the way from barley to beer. For many the presently all-consuming question will be, "Just how many of these do I need to carry out?"

There is no simple answer to this question. The larger brewers will tend to demand a greater spread of analytical measurements, at least in part because of the amplified complexity of their operations. The small guy will be hoping to get by on the bare minimum.

To help address the quandary I believe we must ask ourselves two questions:

- What is and what is not a perceived defect in beer?
- What degree of variation can we tolerate in our product?

Some in the world of brewing science are now asking philosophical questions about whether what is considered anathema by the brewer is necessarily so for the customer. Thus most brewers deplore diacetyl, lightstruck, and cardboard notes on beer. But do customers?

And if we decide that diacetyl, for example, should indeed be low, to what extent can we tolerate a range in its level? Are customers more or less sensitive to diacetyl than is the brewer? When it comes to the plethora of other flavor constituents, how much is the brewer prepared to accept obvious batch-to-batch variations? Will the customer mind or even notice?

The authorities responsible for individual brands must answer these questions. For my part I firmly adhere to the credo of all the major brewers: beer should display minimum batch-to-batch fluctuation in all quality parameters. The following remarks are made with this in mind.

The two overriding rules must be to keep the kit (equipment) clean and to respect your raw materials. Beer is a foodstuff and, frankly, I am appalled at the state of hygiene in many breweries. In short, the whole place should be such as to give your aged aunt a warm feeling of all things being well scrubbed. As for the insides of vessels and pipes, they should be pristine. A properly designed caustic or acid cleaning regime followed by good rinsing and use of a hypochlorite or peracetic acid-based sterilant is critical. The key is more good design and process management (QA) than swabbing and plating (QC). A rapid micro check based on ATP bioluminescence may be warranted, provided the intent is to act on it—if the count is too high, the tank should be re-cleaned.

All conscientious brewers must inspect the raw materials using the most acute instrumentation available to them, namely their senses of sight, smell, taste and touch. (As far as I am aware nobody is listening to their raw materials, but nothing ever surprises me. There are princes who converse with vegetation.) Brewers should taste the water, smell and peruse the malt, rub the hops. Furthermore they should carry on with this throughout the process—worts, dilution waters, and so on. And when it comes to the yeast, look after it. In chapter 10, I talked about ways of checking whether the yeast is alive or dead. But if even this simple procedure is beyond the analytical capabilities of a laboratory, the brewer should at the very least be treating the yeast as one of the family: keep it clean, nurture it, and don't let it stay in undesirable places (i.e., beer post-fermentation). For then it will stay hail and hearty. Sadly, yeast doesn't grow old gracefully.

The critical control of brewing operations hinges on the reliable measurement of temperature, time, mass and volume, with the assessment of specific gravity pretty much making a full set for the small brewer. Weighing-in the correct amount of grist (based on extract information supplied with the grist) and striking with the laid-down volume of water according to the prescribed temperature regime is a must. For wort separation, the measurement of gravity allows the assessment of when to cut off to achieve the target gravity in the kettle. It is logical to ensure that a rolling boil is in place for a sufficient period, and that the system configuration (viz. availability of receiving vessels) is such that wort will not receive insufficient or indeed excessive residence in a kettle or a hot wort receiver.

Many small brewers will pitch on the basis of yeast weight. Specific gravity measurement is still the logical choice for monitoring fermentation. Fermentation should be of a sufficient duration to ensure that vicinal diketones are dealt with. In the absence of a measuring system, this will depend on adherence to good cellar practices, notably the correct pitching of healthy yeast into properly aerated or oxygenated wort, with fermentation at the prescribed temperature, perhaps krausening, possibly the use of a temperature ramp and thence to cold conditioning. Once again we see the overriding value of making sure that the conditions are right (viz. 30.2 °F (−1 °C) for beers that are intended for longer shelf lives) rather than dependence on a myriad of measurements.

Table 12.1 ───────────────────────────

QC Checks on Beer

Parameter	Frequency
Clarity	Every batch
CO_2	Every batch
Foam	Pour – every batch; instrumental (NIBEM recommended) monthly
Haze breakdown	Small pack – total polyphenol and Tannometer monthly
Color	Every batch
Ethanol	Every batch
Apparent Extract (ergo Original Extract)	Monthly
Fermentable Extract	Monthly
Bitterness	Every batch by spectrophotometry; monthly by HPLC
Free amino nitrogen	Monthly
pH	Every batch
Volatile compounds	Monthly
Inorganics (notably iron, copper, sulphate, chloride, nitrate)	Monthly
Sulphur dioxide	Every batch if there are legal requirements to label if above a certain level (i.e., 10 ppm in U.S.); otherwise monthly
Polyphenols	Monthly
O_2	Every batch
N_2 (if used)	Every batch
Microbial status	Monthly
Taste clearance	Every batch
Flavor profile (or trueness to type)	Monthly

More Sophistication

I have described a minimalist regime. Many of the larger brewing concerns will submerge themselves in vast arrays of data on innumerable parameters, formerly in the form of huge ledgers and reams of paper but now handled with flashing and dancing computer screens yelling at too many people near and far. (See Table 12.1 for a recommended regime for beer analysis. The in-process checks will be most or all of those I have described in the various chapters of this book.)

To help ensure that the performance plots mentioned in chapter 3 are as healthy as possible, it is important that the brewer gets genuine QA regimes in place, allied to proper use of an auditing team. Larger brewers will establish such teams from within their staff, including inter-plant teams providing a semi-independent and impartial view of what is going on in sister breweries and suppliers' works. Wise smaller brewers will employ third party auditors.

Auditing teams might approach a job from the perspective of assessing how well a plant is performing against the procedures manual. Alternatively they might hit the brewery in pursuit of enhancing a specific attribute of quality. For instance, if there is a concern about the foam stability of a brewery's beers, a team might be appointed to delve into the process from raw material to beer at point of sale to assess likely problem areas.

Table 12.2 lists the types of checks an auditing team would be expected to be making.

A Very Basic Brewing Lab

It is possible to make decent beer with an absolute bare minimum of equipment—there is, after all, the occasional homebrewer who makes beer that is more than respectable!

There is, though, very definitely a positive correlation between the barrelage output of a brewery and the sophistication of its laboratory.

Table 12.2 ━━━━━━━━━━━━━━━━━━━━━━━━━━━━━━━━━━━━
Audit Team Checks: Examples

Process parameter (should be specified in procedures manual)	Check
Cleaning in-place regime	Detergent strength, sequences, swabs (real time microbiology checks – i.e., ATP bioluminescence)
Feedstock storage and handling	Weight and volume control calibrations, cleanliness
Instrument read-outs and control systems	Calibration versus centralized lab. standard
Temperature, times, additions	Performance versus specifications
Vessel condition	Physical appearance
Plant condition	Floors, walls, hoses, pipework, etc.

Quality manuals should be in place to document how operating procedures should be conducted so the team can compare reality with them.

A basic laboratory supporting a sizeable pub or microbrewery and run by a keen and conscientious brewer might have the following inventory. These things are in addition to thermometers capable of accurately measuring the range of temperatures encountered from mash run-off to cold storage. (When wort is boiling, of course, that should be obvious without recourse to measurement.)

Microscope
No self-respecting lab should be without one. Indispensable for quantifying yeast (by hemocytometer), hunting bugs, explaining unsightly bits and pieces in the product, etc.

Friabilimeter
For the most part, the brewer-maltster interaction should be intimate enough for the brewer to be in a position to trust the

analysis accompanying a shipment of malt. The keen brewer, though, may still value an objective view of malt quality. On balance, the best bet is the friabilimeter (www.pfeuffer.com/fria1gb.htm) At a stroke (actually, at a grind), the brewer will get a view of modification. With an additional investment in a 2.2 mm sieve and a balance (the latter of general applicability, i.e., for weighing additions) the brewer will gain a picture of gross under-modification. (See chapter 5.)

Density Meter or Hydrometers

Most decent brewers will be familiar with the hydrometer. Quick and accurate assessment of specific gravity these days, however, is best performed with a digital density meter. It can be used for gauging wort through fermentation and in assessment of the final product.

Spectrophotometer

Few instruments are as valuable as the "spec". Although the spectrophotometric methods for assessing bitterness, VDK's, and color are imperfect, they are nonetheless adequate for basic purposes.

Gases

Equipment for the measurement of carbon dioxide and oxygen in beer is appropriate. Meters are available for each.

Alcohol

Probably the best option nowadays is a devoted alcohol meter. Visit www.orbisphere.com/prod/beverage.htm.

pH

A simple pH meter can be had for a few bucks.

Bug Hunting

ATP bioluminescence by any other name. One supplier is at www.celsis.com. There are others. Just like buying a car, it pays to shop around, at least on the Internet!

A useful starting point for equipment generally would be A. Gusmer Co. (www.agusmer.com/index.html) Another good source is www.morebeer.com.

Bibliography

Methods of Analysis of the American Society of Brewing Chemists, ASBC, St Paul MN, (last update 1999).

Laboratory Methods for Craft Brewers (editor, R.M. Crumplen) ASBC St Paul MN 1997.

Both of the above are available from ASBC at www.scisoc.org/asbc/

Weissler, H.E. "Brewing Calculations" *Handbook of Brewing* (editor, W.A. Hardwick) Marcel Dekker, Inc, New York, 1995.

Lewis, M.J. & Young, T.W. *Brewing*, Chapman & Hall, 1995.

Bamforth, C. *Beer: Tap into the Art and Science of Brewing*, Insight, 1998.

Appendix One ▬▬▬▬▬▬▬▬▬▬▬▬▬
Units

Temperature

In the United States it is still prevalent for brewers to measure temperature on the Fahrenheit scale but most of the rest of the world employs the Celsius measure. Since I consider myself a brewing citizen of the world, I employ the Celsius measure.

$$°Fahrenheit = (°Celsius \times 9/5) + 32$$

Some of the more frequently used temperatures:

	°C	°F
Steeping and germination	14-18	57-65
Kilning	50-110	122-230
Glucanolytic (proteolytic) stands	40-50	104-122
Starch conversion	60-70	140-158
Lautering	72-76	162-169
Boiling	100	212
Fermentation	6-25	43-77
Cold conditioning	-1	30
Pasteurization	62-76	144-169

Volume

1 U.S. barrel = 1.1734 hectoliters (hl)

1 U.S. barrel = 31 U.S. gallons

1 hectoliter = 100 liters (l)
1 U.S. gallon = 128 fluid ounces
l liter = 1 dm^3

Weight
1 metric ton = 1000 kilograms [1 kg = 2.205 pounds (lb.)]
1 ounce = 28.35 grams
1 bushel of barley grain (U.S.) = 48 lb.
1 bushel of barley malt (U.S.) = 34 lb.

Specific Gravity
°Plato = °Brix = %w/w cane sugar
10 °Plato = specific gravity of 1.040

Hops
1 Zentner = 50kg of hops

Yeast
10 million cells / ml approximates to 0.3 kg/hl

Carbon Dioxide
1 vol. CO_2 per vol. of beer = 1.98 g/ liter

Energy
1 Calorie (i.e., 1000 calories or 1 kcal) = 4.18 Joules

Flavor Components
1 ppm = 1 part per million = 1 mg / liter
1 ppb = 1 part per billion = 1 µg / liter
1 ppt = 1 part per trillion = 1 ng / liter

Alcohol
Alcohol content in % v/v (ABV) = 1.26 x alcohol content in % w/v (1 ml of ethanol weighs 0.79g)

The Basics of Malting and Brewing

Appendixes 2 and 3 cover some basics with regard to malting and brewing and also some of the key concepts in chemistry. These chapters are provided for those readers who may find a refresher on these subjects to be of value in digesting the main contents of the book. For a general appreciation of beer and brewing, you might also care to read one of my other books.

The yeast *Saccharomyces cerevisiae* grows on sugar by fermenting it to ethanol, for which a great many people are truly thankful.

If the sugars are derived from the grape, the resultant end product is wine. If the sugars are from apples, the product is cider. If the sugars are from malted barley, the product is beer, the king of beverages (though I am biased).

Barley

Barley starch supplies most of the sugars from which the alcohol is derived in the majority of the world's beers. Some beers are made primarily from cereals such as wheat and sorghum, but in this book I am focusing on the production of beers from malted barley, perhaps supplemented by adjuncts.

Before it can satisfactorily be used for brewing, barley needs to be malted. Malting comprises the controlled germination of barley followed by controlled drying. The raw barley corn or kernel (See Figure A2.1) is hard and not readily milled. Barley

Figure A2.1 ━━━━━━━━━━━━━━━━━━━━━━━━━━━━━━━
Longitudinal Section of a Barley Kernel

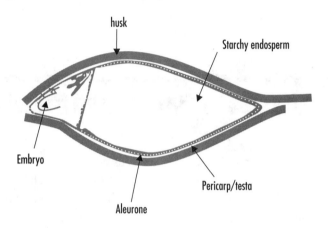

was originally selected for the brewing of beer because it retains a husk on threshing. This forms the filter bed in most breweries in which the extract of malt (wort) is separated from residual material after brewing.

The barley kernel comprises various tissues. The embryo is the baby plant. The starchy endosperm is the store of food that the embryo draws upon to support its growth in the field. To the brewer it is the source of fermentable material that will be converted into beer. The starchy endosperm consists of a mass of dead cells—they don't grow or divide and they don't make the enzymes that are needed to convert their stored components to soluble products the embryo can use.

The cells of the starchy endosperm comprise a thin wall, chiefly of polysaccharides called β-glucans and pentosans, within which are large and small starch granules in a solid sea of protein.

The aleurone is a thin tissue of cells that are capable of vigorous enzyme synthesis. The pericarp, testa and husk are protective layers, the first two being waxy and impermeable to water.

Steeping

The barley corn as harvested is dry (less than 12 percent moisture) or needs to be dried in order to preserve it and help prevent infestation. To enable it to spring into life, its water content needs to be increased (chemical reactions can't occur if the locale is too dry). The target water content is usually between 42 and 46 percent.

Therefore the first stage of malting is steeping—soaking the grain in water. The water enters at a single point in undamaged grain—the micropyle. First the embryo becomes hydrated and then the endosperm.

The structure of the grain is very important in determining the ease with which water is distributed through the starchy endosperm as we can see in chapter 4.

Water uptake and the onset of germination is facilitated if the water is not added in a single batch, but rather interspersed by air rests. The embryo will be starved of oxygen (which it needs to breathe) if it is submerged for prolonged periods. The grain is first steeped for a limited period (about eight hours), before the water is drained and air drawn through the grain. Then after an air rest of about 16 hours, the next steep water is introduced, before another air rest ... and so on. The precise conditions differ between barleys.

Enzyme Development

When the embryo is activated by moisture, it synthesizes hormones. Gibberellins are the most studied of these hormones. These migrate to the aleurone where they switch on enzyme synthesis. The enzymes migrate into the starchy endosperm and break it down (the germination phase has started). The aleurone closest to the embryo is first to be turned on. Degradation (modification) of the starchy endosperm starts at the end nearest the embryo and progresses towards the far end. As the endosperm is degraded it is softened, and the grain becomes more friable. The products of digestion migrate back and are assimilated via the scutellum to support the growth of the embryo. The shoot (acrospire) grows underneath the protective coating and will emerge later in germination from the distal end. The rootlets start to protrude earlier, emerging first as the chit.

Cell Wall Degradation

The first materials that need to be degraded are the cell walls, notably by the β-glucanases. This is important because these glucans are very viscous if they are not broken up and end up being extracted in the brewery where they will cause processing problems.

In a well-modified grain, most of the cell walls have been degraded but there may still be some residual material at the far end of the grain.

Protein Degradation

Roughly 50 percent of the protein in barley is chewed up during the germination phase of malting. The first enzymes to attack are the endo-proteases, 'endo' meaning they chop up the substrate in the middle of the molecule. The peptides that they produce are degraded by the carboxypeptidases which snip off one amino acid at a time: this is from the end with the free carboxyl group, hence their name. They are exo-enzymes, meaning they attack the substrate from the end of the molecule.

Starch Degradation

There are three classes of enzyme responsible for degrading starch. First to attack are the α-amylases. These are endo enzymes that hit in the middle of the molecule with the main products being shorter chains called dextrins.

Next comes the β-amylase. This is an exo-enzyme. It moves along starch and dextrins from one end, chopping off pairs of glucoses. A pair of glucoses joined through the type of bond found in starch is called maltose (malt sugar).

β-Amylase can't get past the sidechains in amylopectin which is that portion of starch that contains branches. These are chopped by a third enzyme called limit dextrinase.

Starch is not particularly easy to digest. In conventional periods of malting, very little starch breaks down, which is a good thing because this is the principle source of fermentable material that the brewer needs. In malting there is some pitting of the large granules, but the small ones largely disappear (which is desirable since they can cause problems for the brewer).

Although there are many more small granules than large ones, quantitatively the large ones hold much more of the starch.

Features of Malting Barley

Not all barley varieties can be malted successfully. So-called malting grades (as opposed to feed grades) degrade their endosperms readily, either because they have an endosperm structure amenable to hydrolysis or because they can make more enzymes.

Barley used for malting should also have a low protein content as, pro rata, for a grain of given size the less protein is present, the more starch there is. Therefore fertilizer use for malting barley should be limited.

Barley also needs to be alive (dead grain can't make hormones) and should also not be dormant. Dormancy is a natural condition designed to prevent grain from germinating when on the ear. The grain is alive but its hormone balance stops it from germinating. Storage relieves the condition. Most types of barley don't need to be stored very long.

Kilning

Kilning is essential to stop germination: by driving off the moisture (to reach a final level in the malt below 5 to 6 percent) the metabolism of the grain is halted, and the malt is stabilized for storage. Kilning must be carried out carefully because the malt enzymes that are needed to carry on polymer degradation in the brewery are heat-sensitive. Therefore heating is started at a relatively low temperature (perhaps 131 °F (55 °C)) to start to drive off moisture. When the water level has perhaps been halved, the enzymes are more stable in this drier environment, and the temperature can be increased.

The other function of kilning is to introduce color and to modify flavor. The more intense the heating, the darker the color. Color is due to the melding of amino acids and sugars. Therefore the better-modified malts, with higher levels of polysaccharide and protein breakdown, will give darker malts during heating. Lager malts are therefore traditionally produced with less modification and less intense kilning.

The flavor changes involve removal of unpleasant characters, such as bean sprouts and grassy notes, and development of pleasing malty aromas.

Mashing

Malting and mashing are really successive stages in a common enzymic conversion of barley into wort. The former is a low temperature process, the latter a high temperature one.

Before mashing, malt is stored for a minimum of two to four weeks. Failure to do this can lead to problems in the brew house. The science underpinning the storage requirement is barely understood.

Milling

Malt must be milled before mashing in order to generate particles small enough to allow ingress of water and egress of pre-formed (in malting) solutes and those produced by enzymic action in the mash. The smaller the particles, the better the solute/solvent transfer. However, traditional wort separation systems depend on the husk as a filter bed, and this must be preserved as much as possible. Mash filters don't have this restriction because filter sheets are used to hold back solids when the wort is run off. Milling for lauter tuns, therefore, involves grinding through rollers with the target of minimizing husk damage but good breakage of the endosperm. Milling for mash filters involves hammer mills designed to comminute the grain substantially.

Continuation of Polymer Degradation in the Mash

The conversion of the polymers begun in the maltings can continue in the mash, depending on the precise conditions therein. The critical parameter is temperature. As we shall see in a moment, mashing of malt must pass through a stage where 143.6 to 149 °F (62 to 65 °C) is attained.

β-Glucanase is a very heat sensitive enzyme and is destroyed in less than five minutes at this sort of temperature. This is a particular problem when a proportion of the malt is replaced by glucan-rich adjunct, i.e., those adjuncts based on unmalted barley or on oats. There are several solutions to this: (a) commence mashing at a relatively low temperature. Many brewers will start a mash at about 122 °F (50 °C), for perhaps 20 minutes to allow the enzyme to survive and act. Then the temperature is ramped to, about, 149 °F (65 °C) at a rate of 1 °C per minute; (b) alternatively brewers may employ decoction processing in which a proportion of the cool mash is removed, boiled, and restored to the main mash with a resultant temperature increase; (c) add heat-stable versions of the necessary enzymes. Various bacteria and fungi produce β-glucanases that are much more heat-tolerant than are the equivalent enzymes in malt.

Endo-proteases are more heat-labile than are the carboxypeptidases. Again, mashing is often commenced at a lower temperature (generally referred to as the proteolytic rest) to facilitate proteolysis. However, many people feel that the extent of protein digestion in a mash is limited, perhaps because there is a range of inhibitor molecules also contributed by malt.

Whereas there are few arguments in favor of restricting cell wall and starch degradation, a balance needs to be struck in the degradation of protein. The amino acids are needed to support yeast metabolism, and proteins that can react with polyphenols to cause haze should be removed. However, some polypeptides need to go through to the finished beer in order to support foam.

The reason a mash needs to be heated substantially is to facilitate starch degradation. As we have seen, granules are relatively resistant to digestion. However if they are gelatinized (which can be likened to melting), then they loosen up and become amenable to attack. For large granules of barley starch this occurs at 143.6 to 149 °F (62 to 65 °C). (Smaller granules have higher gelatinization temperatures and are a problem if they survive malting.) For starches from other cereals (e.g., corn, rice) the temperature is higher, and this is why they are cooked separately before adding to the mash.

Gelatinized starch is amenable to digestion. α-Amylase is a very heat tolerant enzyme, present in abundance, and it will dextrinize the starch from malt and from sizeable quantities of adjunct. β-Amylase, by contrast, is less heat-resistant and will be progressively destroyed in a mash at 149 °F (65 °C). However, it will largely have completed its task of producing maltose in this time, provided it is not excessively diluted out by adjunct.

It is the third enzyme, limit dextrinase, which is the major limiting factor. There are various reasons for this. It is developed later than the other enzymes in malting and therefore is present in restricted quantities. The enzyme that is present is largely bound up with other molecules that restrict its activity. For these reasons, the extent to which limit dextrinase acts in a conventional mash is limited. Perhaps 10 to 20 percent of the starch in a mash is left behind in the form of unfermentable dextrins (The only breakdown products of starch that are fermentable by brewing yeast are glucose, maltose, and maltotriose.). There is a school of thought that these dextrins contribute to the body of beer, but this is unproven. Light beers are those in which the dextrins are fully or largely converted to fermentable forms. This can be achieved by the use of either glucoamylase or pullulanase of microbial origin, enzymes that chop glucose off the dextrins, or by tricks in mashing which allow much more limit dextrinase to act.

The converse requirement is sometimes limited fermentability, in the pursuit of lower alcohol beers. If mashing is commenced at a very high temperature. e.g.,165.2 °F (74 °C), then β-amylase (and limit dextrinase) is rapidly destroyed, whereas α-amylase survives. The resultant wort is very high in dextrins but low in fermentable sugars. The yeast can only take it so far in terms of alcohol production.

Other Enzyme Systems
I haven't yet referred to the enzymolysis of lipids. This is because physical effects are probably more important. Lipids are very insoluble and tend to stick on to the spent grains after mashing. Furthermore, the enzymes involved in breaking them down are not terribly well understood.

The first enzymes of attack are lipases, which progressively split fatty acids off the glycerol backbone. The polyunsaturated fatty acids (i.e., linoleic and linolenic acids) are then substrates for lipoxygenase. Much has been written about the dangers of this enzyme in converting these lipids into the precursors of the compounds that make beer go stale. The arguments are sometimes compelling but generally unproven. Lipoxygenase is certainly a very heat-sensitive enzyme and it is often said that mashing should start at the highest possible temperature to prevent its action. It is also because of lipoxygenase that there is an increasing advocacy of low oxygen ingress into mashes.

Much more problematic may be the peroxidases. There are many heat-resistant ones in malt. These enzymes use hydrogen peroxide to oxidize a wide range of compounds, especially the polyphenols, which originate from the malt, especially the outer layers. The oxidized polyphenols polymerize and produce highly colored entities, which also tend to precipitate out proteins.

Wort Boiling

After perhaps an hour of mashing, the liquid portion of the mash known as sweet wort is recovered (by separation over a time period of two to three hours using a lauter or mash filter) to the kettle (sometimes known as the copper even though they are nowadays more typically stainless steel). Here it is boiled, usually for approximately one hour. Boiling serves various functions, including sterilization of wort, precipitation of proteins (which would otherwise come out of solution in the finished beer and cause cloudiness) and the driving away of unpleasant grainy characters originating in the barley. Because there is a driving off of water, the wort becomes concentrated. This is especially important for those brewers practicing high gravity brewing in which the wort is run to the fermenter at high strength and, after fermentation (in which higher-than-target alcohol levels are produced), the beer is diluted to the desired strength. To facilitate these higher gravities, many brewers also add adjunct sugars at this stage and most introduce at least a proportion of their hops.

The hops have two principal components: resins and essential oils. The resins (so-called α-acids) are changed (isomerized) during boiling to yield iso-α-acids, which provide the bitterness to beer. This process is rather inefficient. Nowadays, hops are often extracted with liquefied carbon dioxide, and the extract is either added to the kettle or extensively isomerized outside the brewery for addition to the finished beer (thereby avoiding losses due to the tendency of bitter substances to stick on to yeast).

The oils are responsible for the hoppy nose on beer. They are very volatile, and if the hops are all added at the start of the boil, all of the aroma will be blown up the chimney (stack). In traditional lager brewing, a proportion of the hops is held back and added towards the end of boiling, allowing the oils to remain in the wort. For obvious reasons, this process is called late hopping. In traditional ale production, a handful of hops is added to the cask at the end of the process, enabling a complex mixture of oils to give a distinctive character. This is called dry hopping. Liquid carbon dioxide can be used to extract oils as well as resins, and these extracts can also be added late in the process to modify beer flavor.

Fermentation and Beyond

After the precipitate produced during boiling (hot break, trub) has been removed by straining through whole hops in a hopback or separation according to centripetal forces in a whirlpool, the hopped wort is cooled and pitched with yeast. There are many strains of brewing yeast, and brewers look after their own strains because of their importance in determining brand identity. Fundamentally brewing yeasts can be divided into ale and lager strains, the former type collecting at the surface of the fermenting wort and the latter settling to the bottom of the fermentation (although this differentiation is becoming blurred with modern fermenters). Both types need a little oxygen to trigger off their metabolism but otherwise the alcoholic fermentation is anaerobic. Ale fermentations are usually complete within a few days at temperatures as high as 68 °F (20 °C), whereas lager fermentations at as low as 42.8 °F (6 °C) can take several weeks. Fermentation is complete when the desired alcohol content has been reached and an unpleasant butterscotch flavor (due to a material called diacetyl), which develops during all fermentations, has been mopped up by yeast. The yeast is harvested for use in the next fermentation.

In traditional ale brewing the beer is now mixed with hops (for dry hop flavor), some priming sugars, and isinglass finings from the swim bladders of certain fish, which settle out the solids in the cask.

In traditional lager brewing the green beer is matured by several weeks of cold storage prior to filtering.

Nowadays many beers, both ales and lagers, receive a relatively short conditioning period after fermentation and before filtration. This conditioning is ideally performed at 30.2 °F (-1 °C) for a minimum of three days, under which conditions more proteins drop out of solution and makes the beer less likely to go cloudy in the package or glass.

Filtration is generally aided by diatomaceous earth (a.k.a. kieselguhr), and various stabilizers, such as polyphenols (removed by PVPP) or proteins (removed by silica hydrogels), may be used to remove materials that may cause haze.

The filtered beer is adjusted to the required carbonation before packaging into cans, kegs, or glass or plastic bottles.

Some Chemistry

Atoms

The basic unit of all matter is the atom.

Modern thinking about the atom is all about waveforms but it is still convenient to talk about the atom in terms of protons, neutrons, and electrons.

At the heart of the atom (the nucleus) are the protons and neutrons, each of which has a mass of one. Protons are positively charged, and neutrons have no charge. Together they comprise the mass of the atom.

Orbiting the nucleus, rather like the planets orbit the sun, are electrons. They are negatively charged but have essentially no mass. In neutral atoms the number of electrons is exactly the same as the number of protons. The atomic weight of an element is its total number of protons and neutrons, i.e., hydrogen has an atomic weight of 1, helium of 4, carbon of 12, and so on.

The electrons orbit the nucleus in defined orbits. There is a limit to how many electrons can occupy each orbit. The orbit nearest to the nucleus can accommodate just two electrons. Because they are so close to the nucleus, the strong positive-negative interaction means that these electrons are less free to move around than those further out—i.e., they have a relatively low energy. The next orbit holds eight electrons. Because they are that much further away they are more energetic. The next orbit holds 18, then the next one 32, then the next one 18 again, then 8, and so on.

Each of the elements in nature consists of atoms. There are well over one hundred elements, each of them having successively one extra proton and therefore one extra electron. The simplest element—hydrogen—has one proton and one electron. The next is helium: it has two protons (and also two neutrons so it has a mass of 4 and not 2) and 2 electrons. And so on.

The most stable (least reactive elements) are those in which the outermost orbit is up to capacity with electrons (2, 8, 18, etc). Helium, then, is very unreactive—and that is why we can be more comfortable flying in airships filled with helium as opposed to hydrogen.

One way in which an atom can complete its outer shell of electrons is by donating electrons to another atom which in turn can complete its outer orbit by accepting electrons. For example sodium can lose a single electron, and chlorine gain an electron, each then assuming full outer orbits. Sodium acquires a net charge of 1+ because it has lost one electron. Chloride gains a net charge of 1− because it has one extra electron. The compound formed is NaCl. The sodium has become a positively charged ion (Na+), sometimes called a cation because it is attracted to a negatively charged electrode (the cathode). Chloride has become a negatively charged ion (Cl−, an anion). The molecular weight of a compound is the sum of the atomic weights of its atoms. So the molecular weight of sodium chloride is 23 + 35.5 = 58.5. The molecular weight of water (H_2O) is $(2 \times 1) + 16 = 18$. A mole is not just a small furry animal: a mole is the molecular weight of a substance expressed in grams. So one mole of hydrogen gas weighs 2 g; one mole of water weighs 18 g, etc.

Magnesium needs to lose two electrons, which it can do by donating one electron to each of two chlorine atoms. Thus magnesium chloride consists of one magnesium and two chlorides, $MgCl_2$. This type of bond is called an ionic bond.

Alternatively an atom can complete its orbital by sharing electrons. Thus if the outer orbitals of two chlorine atoms come into contact, a pair of electrons can be shared between them to make a much more stable molecule, chlorine gas. This type of bond is called a covalent bond.

The number of other atoms that an element can react with is known as its valence. Thus hydrogen reacts with only one atom at a time, therefore valence equals one; oxygen reacts with two to make it more stable, so valence equals two. Ergo we have water with two atoms of hydrogen and one of oxygen, H_2O. Carbon has a valence of four. Sometimes one atom is linked to another by two links—this is called a double bond and is stronger than a single bond. An atom can use up two of its valences in this way Thus in carbon dioxide, CO_2, the carbon uses up its four valences by linking to two oxygens by double bonds, with each oxygen using up its two valences in a double bond to the carbon.

Oxidation and Reduction

When a substance loses electrons, it is said to have been oxidized. The substance that picks up those electrons is said to be reduced.

This is one type of chemical reaction. An example is the conversion of acetaldehyde to ethanol by yeast. Acetaldehyde is reduced, and the molecule $NADH_2$, which is the substance in living organisms that carries the electrons (reducing power), has become oxidized. If one component of a system is oxidized, another component or components must be reduced. The reducing power balances.

$$CH_3CHO + NADH_2 \rightarrow CH_3CH_2OH + NAD \quad (1)$$

$$(\text{n.b. } CH_3CH_2OH \equiv C_2H_5OH)$$

Reactions

All chemical reactions also balance: the total number of atoms of a given element on the left side of a reaction must balance with that on the right. An example would be the conversion of glucose to ethanol and carbon dioxide, which is the overall reaction in alcoholic fermentation of yeast:

$$C_6H_{12}O_6 \rightarrow 2\ C_2H_5OH\ +\ 2\ CO_2 \quad (2)$$

A reaction happens because it is energetically favorable for it to occur. In other words, the total energy of the products is lower (more stable) than that of the reactants. The reaction may not proceed totally to the right hand side: an equilibrium will be established.

This equilibrium may not be established rapidly even if it is thermodynamically favorable. This is because bonds have to be broken in the reactants before new, more stable ones can be formed in the products. A catalyst is a substance that allows chemical species to overcome this energy barrier and speeds up the reaction. The catalyst is left unaltered at the end of the reaction but it may have been temporarily modified during the reaction.

Various other factors influence the rate of chemical reactions. If the reactants are more concentrated, they have an increased opportunity to interact. If the temperature is increased, the molecules collide with greater energy and bonds are broken more readily. (A good rule of thumb, first coined by Arrhenius, is that reactions occur twice as fast for every 10 °C rise in temperature.) And reactions occur much more quickly in more fluid systems. Thus if you mix two powders together in a dry form, they won't react. If they are dissolved, however, the molecules can mix more freely and react together. The solvent often plays a key role in the reaction.

It is possible to group chemical compounds together into families, wherein the molecules have similar structures and similar reactivities. Three such families which are of great importance to the brewer are the proteins the carbohydrates and the lipids. (I'll cover this later.)

It is also possible to make sense out of the complexity of chemistry by realizing that the chemical properties of compounds are determined by the types of groups they contain. There are many types of groupings in chemistry, of which some relevant ones are carboxyl, carbonyl, hydroxyl, and amino. (See Figure A3.1.)

Acids, Bases, pH, Buffers and Salts

The carboxyl group is acidic, in that it can furnish a hydrogen ion (H^+), i.e., a hydrogen atom without its electron, i.e. a proton. (See Figure A3.2.) The amino group can pick up a hydrogen ion and is said to be basic. Another base is the hydroxide ion, OH^-.

The pH value expresses the concentration of the hydrogen ion (H^+) in solution.

$$pH = \log \frac{1}{[H^+]} \quad (3)$$

(Or $pH = -\log [H^+]$, because with logarithms you divide by subtracting, if that doesn't sound too crazy; fellow oldies will remember this.)

Water splits up to a very limited extent to give the hydrogen and hydroxide ions:

$$H_2O \leftrightarrow H^+ + OH^- \quad (4)$$

Any reversible reaction is characterized by an equilibrium constant (K) which quantifies the ratio of concentrations of the various components when the system is in equilibrium (i.e., when the rate of the forward reaction matches that of the reverse direction). In the case of the dissociation of water:

Figure A3.1
Reactive Chemical Groupings

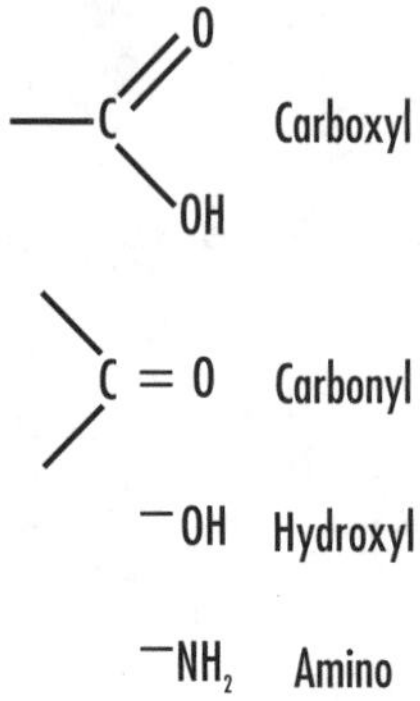

$$-C\overset{\displaystyle O}{\underset{\displaystyle OH}{}}\qquad \text{Carboxyl}$$

$$>C = O\qquad \text{Carbonyl}$$

$$-OH\qquad \text{Hydroxyl}$$

$$-NH_2\qquad \text{Amino}$$

Figure A3.2
Ionization

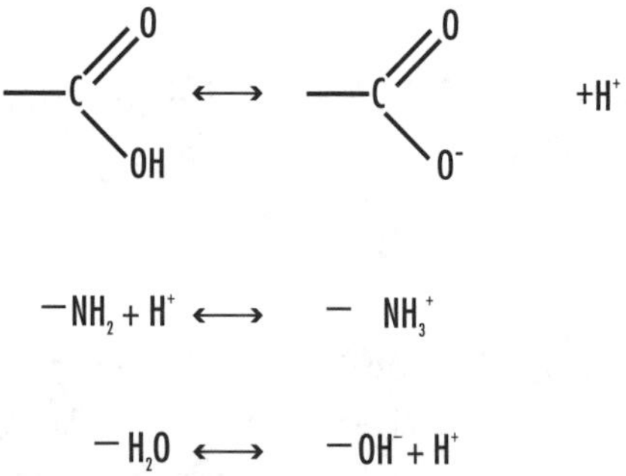

$$-C\overset{\displaystyle O}{\underset{\displaystyle OH}{}}\longleftrightarrow -C\overset{\displaystyle O}{\underset{\displaystyle O^-}{}}\qquad +H^+$$

$$-NH_2 + H^+ \longleftrightarrow\ -NH_3^+$$

$$-H_2O \longleftrightarrow\ -OH^- + H^+$$

$$K = \frac{[H^+][OH^-]}{[H_2O]} \quad (5)$$

This equilibrium constant for water at 77 °F (25 °C) is 1.8 x 10^{-16}. In other words, the majority of water molecules are undissociated.

The concentration of water is 55.5M. (How so? The molar concentration of a solute is defined as the number of moles of that substance dissolved in 1 liter of water. A mole is the molecular weight of a substance expressed in grams. One mole of water is therefore 18 g (2 x 1 g of hydrogen and 1 x 16 g of oxygen). Thus the molar concentration of water is 1000/18 = 55.5M.

So the total concentration of hydrogen and hydroxide ions in neutral water is 1.8 x 10^{-16} multiplied by 55.5, i.e., 1.0 x 10^{-14}M. Since these two ions are present in equal quantities, each is present at 1.0 x 10^{-7}M. This is a very small number, hence the development of the logarithmic pH scale so that the numbers look bigger! When the level of H^+ is the same as that of OH^-, we have a pH of 7, or neutrality. If there is an excess of hydrogen ions, we have a pH below 7, and the solution is acid. If there is an excess of hydroxide ions, the pH is above 7, and the solution is alkaline.

It is important to note that these calculations are based on values taken at 77 °F (25 °C). At 98.6 °F (37 °C) there is more dissociation of water into its ions so neutrality is at pH 6.8. The higher the temperature, the lower this value becomes.

Because pH operates on a logarithmic scale, it will be realized that relatively small changes in pH make for very large differences in hydrogen ion concentration. At 77 °F (25 °C), a drop in pH from 6 to 5 reflects a ten-fold increase in hydrogen ion concentration.

Acids

By definition an acid is a substance that releases hydrogen ions. The stronger the acid, the more readily will it release H +, i.e., the higher is the dissociation constant for the reaction:

$$HA \leftrightarrow H^+ + A^- \quad (6)$$

For acetic acid at 77 °F (25 °C), the dissociation constant is 1.8 x 10^-5. For the much stronger sulfuric acid (i.e., more H^+ released) the value is 1.2 x 10^-2. For lactic acid, which is about 10-fold more acidic than acetic but 100-fold less acidic than sulfuric, K is 1.38 x 10^-4.

Take a 1M solution of lactic acid. Let's say that the concentration at equilibrium of the hydrogen ion (and therefore the lactate anion) is a, then the concentration of undissociated lactic acid must be (1−a).

$$CH_3CH(OH)COOH \rightarrow CH_3CH(OH)COO^- + H^+ \quad (7)$$
$$(1-a) \qquad\qquad\qquad (a) \qquad\qquad (a)$$

Then substituting these values into the expression for lactic acid ionization (25 °C) we get:

$$1.38 \times 10^{-4} = \frac{a^2}{1-a} \quad (8)$$

From this we calculate that a is 0.0022M. Bearing in mind the definition for pH, the pH of a 1M lactic acid solution is 2.66.

Two blokes, one called Henderson and the other Hasselbalch, rearranged the equation depicting the dissociation of weak acids:

$$[H^+] = K_a \frac{[HA]}{[A^-]} \quad (9)$$

where HA is the undissociated acid, A^- is the anion left after the dissociation of H^+, and K_a is the dissociation constant.

If we take logarithms and multiply by −1 (see equation 3), then:

$$- \log [H^+] = - \log K_a - \log \frac{[HA]}{[A^-]} \quad (10)$$

Hence

$$pH = pK_a - \log \frac{[HA]}{[A^-]} \quad (11)$$

where pK_a is $-\log K_a$, the value at which HA and $[A^-]$ are in equal quantities ($\log 1/1 = \log 1 = 0$). The lower is pK_a, the more acidic is HA. Therefore pK_a is a useful index of acid "power"— the lower the value the more acidic is a material.

Buffers

The Henderson-Hasselbalch equation allows us to explain the phenomenon of buffers. Taking equation 6 again, it is obvious that if H^+ is added to the mixture at equilibrium (pK_a) it will react with A^- to form HA, and there will be only a limited accumulation of H^+ (i.e., fall in pH). Conversely if the H^+ present in the equilibrium mixture is removed (e.g. by addition of OH^-, see equation 4), HA will dissociate to release more H^+ so as to restore the equilibrium. Once more the pH change is limited. Consequently an acid-base mixture at its pK comprises a buffer system capable of withstanding changes of pH provided that additions of H^+ or OH^- are not excessive. In fact a buffer operates best within one pH unit either side of its pK_a and is best exactly at its pK_a where the concentration of its acid (HA) and basic (A^-) forms are the same. In other words, if you are seeking to regulate a pH to 5.0, the best buffer to select is one that has its pK_a value in that region. The concentration of the buffering material is also important: the more present, the greater the buffering potential within its buffering range.

Clearly the pH of materials such as wort and beer is determined by the concentration and type of buffer substances present, by the absolute concentration of H^+ and OH^- present or introduced, and by the temperature. Various materials in wort and beer have buffering capacity, notably peptides and polypeptides from the grist (less so the amino acids which tend not to have pK properties in the appropriate pH range).

Factors promoting the level of these materials in wort will elevate the buffering capacity. Such factors will include the nitrogen content of the malt, its degree of modification, and the extent of proteolysis occurring in mashing. As certain adjuncts such as sugars do not contain peptides and polypeptides, their use will tend to lessen the buffering capacity of wort.

The buffering capacity of a wort or beer can be readily assessed by adding acid or alkali to beer and assessing the extent to which measured pH changes.

$$\text{Buffering capacity} = \frac{\text{concentration of } H^+ \text{ or } OH^- \text{ added}}{\text{change in } H^+ \text{ concentration observed}} \quad (12)$$

In turn, the pH impacts on the chemistry happening in all process stages in malting and brewing and the finished beer.

Salts

Let's drift (as it were) back to water which, apart from the grist and the temperature, is the other key factor influencing the pH in mashes. Harder water (which we'll read about momentarily) renders pH's lower than those made with soft water. For example the pH of a wort might be lowered by about 0.4 units by increasing the level of calcium from 50 to 350 ppm. Calcium reacts with carbonate, phosphate, and polypeptides to promote the release of protons and thus lowering of pH. For example:

$$3Ca^{2+} + 2HPO_4^{2-} \rightarrow Ca_3(PO_4)_2 + 2H^+ \quad (13)$$

An important concept encapsulating factors that dictate pH is residual alkalinity of water. Fundamentally residual alkalinity combines in a single term the relative levels of the two key determinants of pH in water, namely the total alkalinity (i.e., level of alkaline substances, notably bicarbonate, as determined by titration) and the level of hardness (as determined from the level of the calcium and magnesium).

$$\text{Residual alkalinity} = (\text{Bicarbonate}) - \left[\frac{\text{calcium}}{3.5} + \frac{\text{magnesium}}{7.0}\right] (14)$$

(The concentrations are quoted as mval's, i.e. milliequivalents per liter. The equivalent weight of a material is its molecular or atomic weight divided by its valence.)

Bicarbonate serves to increase pH, whereas calcium and magnesium lower it through their interactions. The higher the residual alkalinity, the greater the total alkalinity relative to hardness, and so the higher the pH will be. It will be appreciated that two waters might be identical in terms of inherent alkalinity (e.g. the waters of Burton-on-Trent and Munich have very similar bicarbonate levels and therefore alkalinity) but very different in respect of residual alkalinity (Burton water contains far more calcium and magnesium than does the Munich equivalent). Values for the ionic content of various waters around the world are given in Table A3.1.

Table A3.1
Ionic Composition (mg per liter) of Water

Component	Burton	Pilsen	Dublin	Munich	Davis
Calcium	352	7	119	80	40
Magnesium	24	2	4	18	90
Sulfate	820	5	54	8	60
Chloride	16	5	19	1	50
Bicarbonate	320	37	319	333	450

Hardness

Hard water is water that does not lather easily with soap, whereas soft water does. The difference is due to the presence of many more salts in hard water. Hardness may be of two types: permanent or temporary.

Permanent hardness is caused by the sulfates (and perhaps chlorides) of calcium and magnesium. When water is boiled these sulfates don't change, save that they get more concentrated as water evaporates. Such water is found in regions of high gypsum content.

Temporary hardness is due to the bicarbonates of calcium and magnesium. This water is found in limestone, chalk, or dolomite regions. If these solutions are boiled, the bicarbonate decomposes, releasing carbon dioxide, and reducing the hardness. Addition of lime (calcium hydroxide) has a similar effect. Acid may be added to reduce alkalinity and temporary hardness. If the water is rich in calcium bicarbonate, addition of sulfuric acid will convert temporary into permanent hardness. You'll notice in Table A3.1 what we in Davis must contend with—temporary hardness big time.

Polymers

The most significant classes of compound we need to address when considering malting and brewing are the carbohydrates, proteins, and lipids. These are polymers: a cell stores materials in this way to avoid osmotic stress.

Carbohydrates

Many of these compounds have the general formula $C_n(H_2O)_n$, i.e., are hydrates of carbon (hydrate as in hydration—addition of water)—thus the word carbohydrate. So, glucose has the simple formula $C_6H_{12}O_6$.

In fact glucose has a more complex structure than this (See Figure A3.3). For the most part it is found in a ring form. Two

Figure A3.3 ━━━━━━━━━━━━━━━━━━━━━━━━━━━━━━━━━━━━━━

The Structure of Glucose

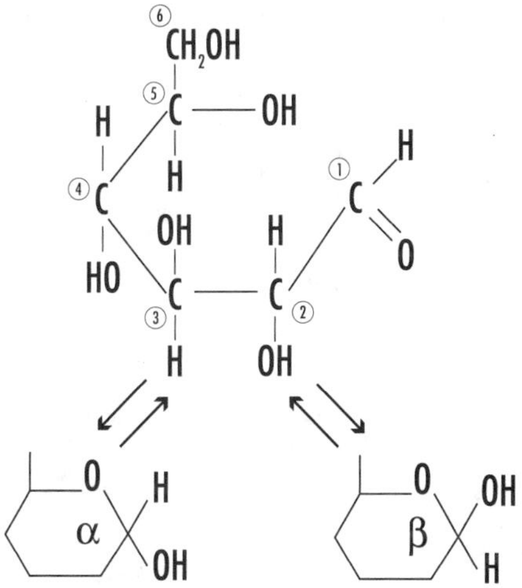

ring forms are interchangeable through a linear form. These two forms are called α and β. Carbon atom number one, is also called a reducing group because it has a free carbonyl group, which is readily oxidized (and as we have seen, when something is oxidized something else is reduced).

Low molecular weight carbohydrates tend to be freely soluble in water and sweet. They are called sugars. For many organisms, including the barley embryo and yeast, sugars are the primary source of fuel that is used for burning to generate energy.

Most of the carbohydrate in a barley kernel is polymerized with the glucoses joining together with the splitting out of water (condensation; the reverse process where water is added to break these links apart is called hydrolysis). The two principle polymers of glucose are starch (α-glucan) and β-glucan, which is the principle component of the walls surrounding the cells in the

Figure A3.4 ──

The Structure of Starch

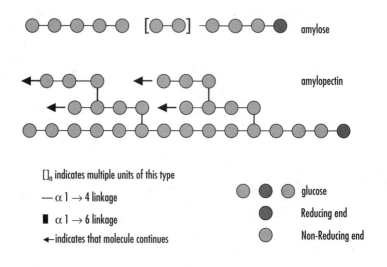

starchy endosperm. The starch is in the form of granules that are packed inside these cells.

When adjacent glucoses link with the glucose in the α-conformation, we get an α-glucan, i.e., starch. (See Figure A3.4) There are two types of molecule in starch: amylose with linear chains of glucose units linked α1-4; and amylopectin with additional side-chains, in which glucoses are linked α1-6. In both cases the last glucose in line (the one in which its C-1 is not attached to another glucose) is called the reducing end. The other end (one per molecule in amylose, many per molecule in amylopectin) is called the non-reducing end.

When adjacent glucoses link with glucose in the β-conformation we get β-glucans. In the cell wall glucan of barley 70 percent of the glucoses are linked β1-4, and 30% β1-3. To a first approximation every third or fourth link is a β1-3.

Proteins

The other major component inside these cells is protein. Again this is a polymer but the monomers here are the amino acids.

Amino acids have the general formula shown in Figure A3.5. There are approximately 20 amino acids, which differ in their R group. (See Figure A3.5.) The simplest, glycine, has just an H atom. Others have carboxyl groups, amino groups, or hydrophobic residues, etc.

The amino acids join together through peptide bonds, again with the splitting out of water. Arbitrarily we can say that a peptide contains up to 10 amino acids, and a polypeptide rather more. A native protein can be very large or very small. Some consist of several polypeptide chains. I reserve the word protein for any undegraded protein which still has its functionality intact, i.e., it is an undegraded storage or structural protein or an enzyme. (I'll cover this later.) The word polypeptide is restricted for a part of a protein or a partially-degraded protein.

Figure A3.5 ━━━━━━━━━━━━━━━━━━━━━━━━━━━━
The General Structure of An Amino Acid

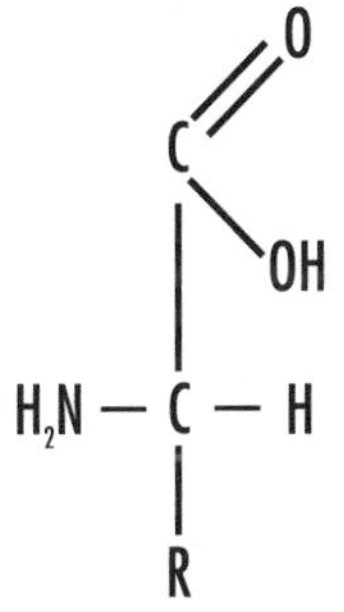

The side chains of the amino acids determine the properties of a protein. For example, negatively charged ionized carboxyl groups can react with adjacent positively charged amino groups, and this will influence the shape of the protein. The hydrophobic groups tend to congregate in the middle of the protein, away from the water which solvates the outside of the protein.

Lipids

There are various types of lipids in nature, united by the common feature that they are insoluble in water (c.f. oils and fats). They are usually found in the membranes that surround cells (e.g., the cells in the embryo of barley), but some are associated with starch. The most important of the lipids consist of fatty acids esterified with glycerol. It is the long chains of the

Figure A3.6
The Active Site In An Enzyme

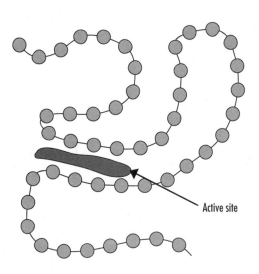

Active site

fatty acids that makes them insoluble in water. There are several of these fatty acids, those receiving most attention being the unsaturated fatty acids, which are readily oxidized.

Enzymes

Enzymes are biological catalysts. We have already seen that catalysts speed up chemical reactions without themselves being modified at the end of the reaction. Enzymes are nature's catalysts, allowing living organisms such as barley and yeast to perform their functions at moderate temperatures.

Enzymes are mostly proteinaceous in nature. The reactions they catalyze occur at a given location on the enzyme called the active site. (See Figure A3.6) This site may involve amino acids from different parts on a polypeptide chain (or even on different polypeptide chains). These amino acids are brought together because of the complex folding of the molecule.

Anything that disrupts this folding will also disrupt the active site and therefore inactivate the enzyme. Disruptive factors include heat and shifts of pH. This is why many enzymes are destroyed at relatively low temperatures and also why most enzymes are only active within a specific range of pH values.

The substance that the enzyme acts on (substrate) binds to the active site. The more enzyme present, the more active sites to bind to, therefore the faster reaction. The more substrate, the more active sites will be filled and the faster the reaction—but only up to a point. When all of the active sites are filled the enzyme is said to be saturated. Reaction rate is at a maximum.

For enzymes we usually talk of two parameters, the K_m and the V_{max}. (See Figure A3.7.) The former is the Michaelis constant and is the substrate concentration at which half maximum reaction rate is obtained. The lower this value, the greater is the affinity between enzyme and substrate. V_{max} is the reaction rate observed when all the active sites are filled, and the enzyme is operating at its maximal rate.

Figure A3.7 ━━━━━━━━━━━━━━━━━━━━━━━━━━━━

Basic Enzyme Kinetics

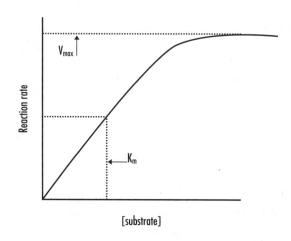

[substrate]

K_m = substrate concentration giving half
maximum velocity (V_{max})

Various factors influence the rate at which an enzymic reaction will occur: (a) substrate concentration; (b) enzyme concentration—the more enzyme molecules, the faster the reaction; (c) temperature (There is a balance between the effect of heat in speeding up chemical reactions and heat also inactivating enzymes, see Figure A3.8); pH (because of the effect on the side chains in the amino acids in the enzyme, which in turn will influence the 3-D structure of the enzyme but also the charge of groups that may be involved in the catalytic act in the active site).

Enzymes can also be inhibited, for example, by materials that are able to react with the active site but that can't be broken down (these are competitive inhibitors) or by materials which poison groups on the enzyme, thereby disrupting the delicate structure of the protein. (These are sometimes called inactivators, and they include heavy metals such as copper.)

Figure A3.8 ━━━━━━━━━━━━━━━━━━━━━━━━━━━━━━
The Impact Of Heat On An Enzyme-Catalyzed Reaction

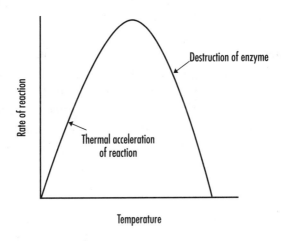

Temperature

───

Some Exercises on Water and pH

1. What is the pH of an unbuffered solution containing (a) 0.0001M hydrogen ions; (b) 0.0000000003M hydrogen ions?

2. What is the concentration of hydrogen ions in unbuffered solutions of pH (a) 5.29 (b) 11.12?

3. Here are some analytical measurements made on some samples of water:

Mg/L	A	B	C	D
Calcium	11	410	130	90
Magnesium	10	18	6	18
Sulfate	5	796	60	7
Chloride	4	17	21	7
Bicarbonate	43	180	307	380

a. What is the residual alkalinity of these waters?
b. Which of these waters might be expected to lead to the lowest wort pH?
c. Which water would need most acid to lower its pH?
d. Which of these waters would be softened most by boiling?
e. Which of these waters is most likely to be used for the brewing of a lager-style beer?
f. What would be the impact of adding sulfuric acid to water D?
g. Which of these waters might be expected to leach most polyphenol if used untreated as a sparge water?

Some Common Laboratory Practices

I remember visiting a brewer in the northwest of England once and asking him to show me his lab. He smiled and opened a drawer. Within were two items in portable cases: the first was a meter for measuring alcohol, the second was an ATP-bioluminescence kit for checking the hygiene of his plant. Compare and contrast: another time I was the guest of a famous Japanese brewing company who showed me their research facility. Two stories, room after room, stacked with many thousands of dollars worth of automated equipment—even a karaoke room!

It stands repetition so I'll write it again: the most sensitive tools available to brewers are their eyes, nose, and mouth. These are the very sensors our customers will use to assess the quality of beer, and hence the raw materials, and the process stream. Fingers are also valuable to the brewer for the evaluation of malt modification and hop aroma.

Next we need a thermometer, a facility to measure weights and volumes, and a hydrometer for monitoring fermentations After this, it is safe to say that the two most valuable and flexible instruments that should be present in any basically established brewing QA lab in priority order are, first, a spectrophotometer and, second, a chromatograph.

Spectrophotometry

Many chemical compounds absorb light at specific wavelengths. For instance, materials that are yellow appear that color because they absorb light at lower wavelengths (the blue end of the spectrum—see Figure 5.1 in chapter 5). The longer wavelengths, notably in the yellow region, are not absorbed so they pass into our eyes making what we see the color yellow. Something that looks blue to us does so because it absorbs light at longer wavelengths.

The more highly concentrated a substance, the more light it absorbs. So by measuring light absorbance at specific wavelengths, we can measure how much of a specific material is present. Sometimes we have to extract the substance first. We've seen how beer is extracted with iso-octane before measuring bitter substances via the amount of ultra-violet light they absorb at 275nm. Often the substance doesn't itself absorb light in a useful way but will after reacting with something else. Thus diacetyl makes a colored derivative with α-naphthol. Another trick is to use an enzymic reaction to measure a compound through the absorbance of something else. An example is the measurement of alcohol. We have already encountered the enzyme alcohol dehydrogenase—the enzyme that's responsible for turning alcohol in yeast. It also works in the opposite direction:

$$CH_3CH_2OH + NAD \rightarrow CH_3CHO + NADH + H^+$$

If the pure enzyme (readily available from suppliers) and NAD are mixed with beer in a buffered mixture, the alcohol is oxidized and NADH is produced. NADH absorbs u.v. light at 340 nm whereas NAD doesn't. Thus by measuring the increase in absorbance at this wavelength you can get a measure of NADH. Each NADH equates to one ethanol molecule, so you can have a measure of alcohol.

I hope you have a feeling for the flexibility of spectrophotometry.

Chromatography

Another useful technique for measuring materials is chromatography. This involves separation of mixtures in columns between a moving phase and a stationary phase. If a substance has more affinity for the mobile phase as opposed to the stationary phase, it moves a long way down the column. If it prefers the stationary phase, it moves to a lesser extent. The individual substances are detected, perhaps by measuring their absorption of specific wavelengths of light, staining them with a dye, assessing their ability to conduct an electric current, and so on. In gas chromatography the mobile phase is a gas and the stationary phase some type of high surface area solid. It is widely used for measuring volatile flavor active materials: In High Performance Liquid Chromatography the mobile phase is a liquid at very high pressures. It can be used for measuring non-volatile materials, such as sugars, bitter acids, and polyphenols.

Answers to the Exercises

Chapter Five

1. a. B
   b. C
   c. C
   d. A
   e. A
   f. B
   g. B
   h. B
   i. The barleys could have been separated through sieves of different hole sizes.
   j. C

2. 10.75 %

3. A, 4.8mg; B, 8.2mg; C, 4.5mg

4. a. Prisma (predominates) and Halcyon; b. Pipkin (predominates) with some Fanfare

5. Barley B would be rejected on account of its high DON content. It also has a very high nitrogen (protein) and β-glucan content. Barley A is very dormant (difference between Germinative Capacity and Germinative Energy), but otherwise

has pretty good malting properties. Barley C contains 10% dead grain, all the viable ones being non-dormant. It has bigger kernels than A and the viable ones will modify better than those in A because they are more mellow. The 10% dead grain is a result of pre-germination on the ear.

6. Detailed interpretation of the data would be performed with aid of Youden plots. Lab b measuring high for nitrogen. Method for β-glucan clearly lacks robustness as there is much scatter between labs, with reasonable agreement only between HQ and lab c.

Chapter Six
1. a. D
   b. Friability, acrospire length, fine coarse difference in extract. (To explain the last of these: if the malt is milled very finely before measuring extract then the value obtained will be similar between malts of different modification because the fine milling will equalize differences in modification. If the malt is milled coarsely, then only the well-modified kernels will release their extract readily. The difference between extract values determined after fine and coarse milling is therefore an index of modification.)
   c. F
   d. E
   e. F
   f. D
   g. D
   h. D
   i. D
   j. D

2. a. B (will have higher β-glucan after correcting for moisture content)

b. Both the same [as we are dealing with a ratio (soluble nitrogen ratio) this will not differ depending on moisture content]

3. A. Crystal
   B. Cara Pils
   C. Black
   D. Chocolate
   E. Amber
   F. Munich
   G. Maize grits
   H. Torrefied barley
   I. Flaked rice

Chapter Eight

   a. A
   b. A
   c. D
   d. B
   e. C
   f. B

Chapter Nine

1. a. F
   b. 13.525
   c. G
   d. G
   e. E
   f. F

2. 17 grams

3. 71.9 °C

4. 15.4 tons malt, 4 tons syrup

5. 35 hL

6. 27.4%

Chapter Ten

1. The cells in the second slurry are (on average) smaller

2. Number

3. A, lager; B, ale; C, mixed; D, lager

4. B is the standard fermentation; In A there has been insufficient aeration/oxygenation of the wort or perhaps there is a zinc deficiency; In C extra enzyme (glucoamylase) has been added.

5. C

7. 309 liters

Chapter Eleven
   a. H and I
   b. I will likely have been fermented with an addition of glucoamylase
   c. Probably I, though an increase by base addition
   d. H – lower CO_2 content
   e. H – because it's from same wort as I
   f. Post-fermentation bittering
   g. I
   h. High iron
   i. J
   j. J

k. H and I similar
l. Sulfite on beer J
m. H, 40.8; I, 41.9; J, 44.7

Appendix Three
1. a. 4.00
   b. 9.52

2. 5.13 x 10^{-6} M b) 7.59 x 10^{-12} M

3. a. First the values for bicarbonate, calcium and magnesium
      need to be converted into mval. Taking the bicarbonate level
      in water A as an example. The molecular weight of
      bicarbonate (HCO_3) is $1 + 12 + (3 \times 16) = 61$. The valence
      of bicarbonate is 1, so its molecular weight and equivalent
      weight are the same. Therefore 1 mval of bicarbonate is 61
      mg. Therefore 43 mg/L is 0.7 mval. (Note the atomic weights
      of calcium and magnesium are 40 and 24 respectively and
      both have a valence of 2.) So to continue with water A, we
      have 0.14 mval of calcium and 0.21 mval of magnesium.
      Using the equation in the text, then, the residual alkalinity is
      $0.7 - (0.14/3.5 + 0.21/7.0) = 0.63$ mval/L. For the other
      waters the values are B, 1.44; C, 4.55; D, 5.86.
   b. B
   c. D
   d. D
   e. A
   f. Convert temporary to permanent hardness
   g. D

Index

References to figures are printed in **boldface** type. References in *italics* refer to tables.